THE

MAGNETISM OF SHIPS,

AND

The Mariner's Compass;

BEING

A RUDIMENTARY EXPOSITION OF THE INDUCED MAG-
NETISM OF IRON IN SEA-GOING VESSELS,

AND

ITS ACTION ON THE COMPASS, IN DIFFERENT LATI-
TUDES, & UNDER DIVERSIFIED CIRCUMSTANCES.

BY

WILLIAM WALKER,

COMMANDER, R.N.,
EXAMINER IN NAVIGATION AND SEAMANSHIP, ETC.

SECOND EDITION, REVISED.

DEVONPORT:
H. V. HARRIS, PRINTER & STATIONER TO HER MAJESTY,
15, FORE STREET.

1863.

PREFACE.

WHEN a person writes anything in the shape of a book for publication, he is expected to give a reason why he writes it, and for what purpose. In order to satisfy this expectation, I have to inform the reader that, although cheap elementary and popular treatises on almost every subject relating to mechanical, manual, or practical industry can be obtained in a condensed form, and at small cost, one looks in vain for a sound, practical, elementary, and popular work on the induced Magnetism of Ships, and the influence it exerts on the Mariner's Compass.

A work of this kind, if written in language familiar to seamen, and yet sufficiently comprehensive in detail, would necessarily tend to expand the mind of a young navigator. Parliamentary enquiries have shewn, that in every week there are at least *ten* British merchant ships wrecked, stranded, or seriously damaged; and ships are constantly running in one direction on a *compass course*, whilst the persons in charge firmly believe they are going in another. Ships shape compass courses towards rocks and shoals, by orders from the navigators, who think they are perfectly safe in shaping such courses. It is in this way that much property and many lives are lost, and no small amount of professional reputation called in question. The writer, therefore, thinks that, if more were known about the Magnetism of Ships and the Mariner's Compass, fewer misfortunes would befall the shipping of this and of other countries.

Being sensitively alive to nautical matters, and impressed with an opinion that sufficient skill and precaution were not employed at sea, he wrote, some ten or twelve years ago, several articles on these subjects for the periodicals of the day; such as the *Naval and Military Gazette*, the *Shipping Gazette*, and *Nautical Magazine*; and some of his remarks have been quoted, or have found places in works of a more permanent character.

Having acquired much personal experience at sea, as a navigator, in the mercantile marine, as a master in the Royal Navy, as an officer in the Royal Dockyards,—having had the superintendence of all kinds of ships and compasses in the Royal Navy, and in swinging them for compass deviations,—it was considered, that by bringing together some of his previously published remarks, notes, hints, and practical experiments, to form a small volume, might prove of some practical utility to young seamen aspiring to rise in their profession.

Seamen, however, are generally persons with a very limited amount of what is usually called *education*. They are generally persons from the humble walks of life, who have been sent to sea at a very early age, in order to be inured to a life of danger, toil, and privation. The young sailor has a vast amount of knowledge to acquire before he can be justly entitled to the designation and rating of A.B. or able seamen. An able seamen, properly so called, has all his practical duties at his fingers' ends, and knows exactly what to do, and how to do it; but his limited amount of education disqualifies him for graphically describing the principles of his art to another. However gifted he may be in the actual practice of his calling, or in the intuitive performance of seamanship, he cannot communicate, in writing, to others the knowledge which he acquired by experience alone. Hence it has resulted, that almost all our practical and scientific works, on Navigation and the Management of Ships, have been produced, compiled, or methodically arranged, by landsmen; who, however gifted in other respects, would be found unfit to put into practice at sea the principles and precepts they undertake to propound in books.

The young sailor's earliest years are passed in watching or working on deck or aloft by day, and in making and shortening sail or looking out at night. The place appropriated for his relaxation or repose, in the forecastle of a merchant ship, is often dark, ill-ventilated, and unsuitable for a place of study or mental improvement. A sea boy, then, has but slender means for acquiring the necessary amount of education which every person should possess who may be entrusted with even the charge of a watch in a ship at sea; and yet it is to these boys that we must necessarily look for all the officers of our extensive and valuable mercantile marine.

Iron now enters largely into the formation of a ship's hull, rigging, or cargo. The compass, often the helmsman's only guide, can seldom be placed beyond the influence of the induced magnetic action of the iron in a vessel.

Coast navigation, in dark nights, or in foggy weather, has become more hazardous. Sensible of these facts, the Lords Commissioners of the Admiralty now require, that young officers, under examination for navigation at the Royal Naval College, shall possess a certain amount of knowledge relative to a ship's local attraction.

The Board of Trade, under authority of the Mercantile Marine Act, inserts, in the examination sheets for masters and mates in the mercantile marine, certain nautical problems for candidates to solve; and among other things they are required to understand, and to "explain the meaning of deviation of the compass, and the method of determining and correcting it." I have however found, that it is *instruction*, rather than examination, that these candidates require. How can it be otherwise, when not one in twenty of those who undertake to teach *navigation* know anything more of the action of iron on a ship's compass, than that it is liable to attract and derange the free action of the needle?

Navigation is now taught in an instructive and efficient manner by *able teachers*, in the nautical branch of the Upper School, in the

Royal Hospital at Greenwich.* I was recently fortunate in being present at a half-yearly examination of the boys, and was much gratified by finding that they are now regularly taught the known principles of a ship's attraction on its compass under different conditions,—principles which a few years ago I ventured to submit to the Admiralty and Compass Committee, and which, at that time were deemed rather visionary, or of small practical importance. These principles are, however, now regarded by government as a necessary part of a navigator's education, and he is legally required to obtain this knowledge where he best can find it.

Having now shewn why this unpretending volume was put together—for what purpose it was intended—that a necessity exists for an elementary treatise of the kind, in order that seamen might better understand the subject,—it is now set adrift, to be picked up by such persons as may require practical information on the Magnetism of Ships, and the Mariner's Compass.

W. W.

Plymouth, June, 1853.

PREFACE TO THE SECOND EDITION.

The first Edition being out of print, and a second being called for, the author has revised and condensed his book, and has expunged some passages, in order that this Edition may be sold *for less than one-third of the price of the first Edition!!* Iron ships, wooden ships with iron cargos, and wooden or iron ships of war defended by iron sides, decks, and towers, have made navigation more difficult under compass guidance. The present Edition is therefore a work of necessity.

W. W.

Plymouth, 1868.

* The Board of Trade and Committee of Council for Education, now promote navigation Schools, conducted under properly certificated teachers, and Managements of Local Marine Boards, assisted by capitation grants. (1869.)

CONTENTS.

CONTENTS.

CONTENTS.

ON THE MAGNETISM OF SHIPS, AND THE MARINER'S COMPASS.

§ 1. IF our knowledge of the properties of the loadstone had still been confined to its power of attracting iron, we must have remained ignorant of the form and magnitude of the earth, of its proportions of land and water, and of the various races of men and other animals that inhabit it. The mariner must still have continued to row or sail slowly along the land during fine weather, with a fair wind and a clear sky; and if he ventured beyond the ordinary limits of his knowledge, it would behove him to look out for a place where he might "beach his boat," or secure her in some sheltered creek, before darkness or foggy weather should overtake him. But the magnetic properties of the loadstone were ordained for useful purposes; for although the discoveries of these properties were made but slowly, and even yet remain covered, as it were, by a semi-transparent veil, their practical utility has been very great.

The introduction of the mariner's compass, even in its primitive and rude state, brought about a complete revolution in the theory and practice of naval architecture and seamanship. It extended our geographical knowledge, and opened a social and commercial intercourse between different nations of the earth. By it countries previously unknown to Europeans were discovered and colonized; and the ends of the earth were actually joined together by circumnavigation.

In February, 1817, the author was appointed to command H.M. Store Ship *Despatch*, to be employed in the conveyance of naval timber from a forest on the south-east coast of Africa, to the Dockyard at Simon's Town, Cape of Good Hope. In this vessel he made fourteen trips round Cape Aguilhas,* a name, by-the-by, given to this Cape by the Portugese, on account of the magnetic derangement of their compass needles in its locality when Vasco de Gama sailed round it.

"Aguilhas" in Portuguese signifies Needles.

B

The *Despatch* had about 40 tons of cast iron ballast, chain cables, two guns, and wrought-iron diagonal riders in her hold. In coasting along in an east and west direction, the vessel was constantly getting to the northward of her reckoning, or farther to the north-ward than she should get. On applying a pocket compass to the spindle of the capstan or other article of iron, it was evident the local magnetism of the vessel had changed since she left our northern latitude. On returning to England, the ship was again found to deviate from her reckoning in a southerly direction, and not in a *northerly*, as was found to be the case at the Cape of Good Hope.

In the meantime iron tanks were introduced, chain cables and iron knees, and a Mr. Bill took out a patent for iron bowsprits and lower masts for men-of-war. The Lords Commissioners of the Admiralty ordered H.M.S. *Phæton* to be supplied with an iron bowsprit and a mainmast on Mr. Bill's plan. I wrote a letter to Admiral Sir Byam Martin, then Comptroller of the Navy, stating the result of my experiments on the magnetism of ships; and I gave an opinion that the heels of iron lower-masts and bowsprits would repel the north point of the compass needle, and I recommended that the effect should be observed in the *Phæton*. I received the following reply to my communication :—

"Buckland, Ashburton, October, 19th, 1822.

"SIR,—I am much obliged to you for the judicious observations contained in your letter of the 4th instant, in reference to the effect of iron masts on the compass; and you may be assured that a matter of so much moment will be carefully watched.

"I am, Sir,
"Your very humble Servant,
"Mr. W. Walker."   "T. B. MARTIN."

It appears, by an able article in the first volume of *Papers on Naval Architecture*, p. 100,* that Admiral Martin requested Pro-fessor Barlow to report upon the effect that iron masts would pro-duce on the *Phæton's* compasses; that on swinging the ship her greatest deviation did not exceed 1° 40′—an error unusually small in a frigate of her class; that the heel of the iron mainmast repelled the north point of the compass needle nearly as much as the whole of the guns, tanks, iron ballast, and shot of the vessel attracted the north point; and that when the iron mast was ultimately taken out and laid upon the ground in Portsmouth Dock-yard, its head and heel retained a portion of the magnetic polarity it had acquired in its vertical position when "stepped" in the ship. The statements I had made to Sir T. B. Martin, in my letter of the 4th October, 1822, had consequently been verified.

The mariner's compass is held in veneration by a thorough sailor. In a dark and cloudy night, or during a thick fog, he steals softly aft under pretence of putting something to rights; but his object is to take a glance at the compass to see how the ship's "head

* Observations on the effect produced by Iron Masts, &c. on the Compass Needle; with an Account of an Experiment made to ascertain the Local Attraction of H.M.S. Phaeton (Capt. Hurd), fitted with an Iron Mainmast and Bowsprit. By Mr. J. Bennett, Naval Architect,

lies." Seamen know but little of the doctrines of magnetism, but they know full well that the compass is their only guide at sea, and that by it the ship's course is shaped. If a sailor discover an iron nail, or a marline spike, left by some "lubber" near the binnacle, he slily consigns it to "Davy Jones' locker,"* without any qualms of conscience, for he knows intuitively that iron has no business there.

To those seamen who are navigators, it is considered that a brief essay on the mariner's compass will not only be amusing, but really useful in their profession. It is my opinion, as an experienced sea-man, that if *more* were known by navigators of practical magnetism, (as for example, how the stowage of a ship's cargo, or the arrange-ment of the iron within a ship, might affect her compass,) *fewer* ships would be lost; for all those ships that actually *run on shore with a fair wind*, when steering a compass course *intended* to lead them clear of all danger, are without doubt *wrecked* through want of skill in the navigator and a knowledge of his compass deviation.

It is, therefore, my intention to present the reader with a con-densed account of the mariner's compass, and of the very slow progress that practical magnetism has made; and how this know-ledge has been applied to purposes on land, as well as at sea. I shall give a short notice of the theoretical views that have been, from time to time, entertained by philosophers of magnetism; and shall conclude by shewing the *practical application* of what is really known of the principles of local attraction, and in what way a ship's reckoning is liable to be influenced by the local magnetism of a ship and her contents.

§ 2. Although the Greeks, Egyptians, Phœnicians, Carthagenians, and Romans, had ships fitted for coast navigation, and generally capable of entering shallow waters, or of being hauled on shore, they have left us no historical record of anything like a compass being used in their vessels. They knew so very little of the magnet and its properties, that their priests had not attempted to impose the mysteries of magnetism on the credulity of the people. All they knew of the magnet was, that iron was attracted by it. From the days of Homer to the time of the Crusades, in the 12th century, there are good grounds for believing that the magnet was not in any way applied to purposes of navigation in Europe.

The Chinese are, without doubt, a very ancient people; and although I am not disposed to believe *all* that has been translated to us from their history, yet a good deal of the Chinese history has been *connected with the history of the Heavens;* and, therefore, verified to a certain extent. The Reverend Pere Ganbil examined the records of thirty-six eclipses of the sun in the Chinese history, and found only two doubtful and two false. The author of the *Histoire Universelle*, in speaking of China, says, "La Boussole ainsi que la Poûdre à tirer étoit pour eux une simple curiosité." And in another place, "La Boussole, qu'ils connoissoient, ne servoit pas à son veritable usage de guider la route des Vaisseaux; ils ne navi-

* That is, he throws it overboard.

goient que près des côtés." We are informed by Du Halde, who was a missionary in China, and who wrote a history of China from data he procured from Chinese books, that about the year 2634 B.C. the Emperor Hoang-ti, being at war, an instrument was invented, which, being placed in a *car*, *pointed to the south*, and enabled the imperial army to direct its march, and surprise the enemy during a thick fog. If this statement be correct, it affords evidence of the Chinese making use of the directive power of the magnet 4497 years ago. The same author informs us, that 2893 years ago an embassy reached China from Cochin; that the ambassadors had experienced great difficulty in finding their way to the imperial court; but on taking their final audience, Du Halde says, "Tcheou-kong gave them an instrument, of which one end pointed to the *north* and the other to the *south*, that they might find their way home with less embarrassment than they had experienced in their route to his dominions. The instrument was then called Tchi-nan, and this is the name which the Chinese now give to the mariner's compass."

§ 3. In a Chinese Dictionary, compiled about the end of the fourth century, there is the following passage: "They had then ships which directed their course to the south by the magnetized needle."* "The *fortune-tellers* rub the point of a needle with the *stone of love*, for rendering it proper to indicate the south." These extracts shew that the Chinese made use of magnetism for land and sea voyages, at a very early period of their history. The old Venetian traveller, Marco Polo, whilst in the service of Koublia Khan, obtained the command of a Chinese fleet of fourteen ships, each ship having *four masts and nine sails*. This fleet was prepared to convey a Chinese Princess to the Persian Gulf; it sailed from the river of Pekin early in the year 1291, and was eighteen months in making a passage to the Persian Gulf. Some of these junks had crews of 250 men. This expedition is mentioned, in order to shew that, in former times, the Chinese undertook longer sea voyages than they now undertake. It is extremely probable that the compass was in common use in the fleet referred to, although no mention is made of it in Marco Polo's Narrative.

The compass of the modern Chinese is probably nearly similar to those used two thousand years ago. One in the writer's possession may be thus described: A very small steel bar, about an inch in length and of the diameter of a sewing-needle, poised with great correctness, and *strapped* to the top of a small copper hemispherical cup, which serves as a socket to receive the point of a vertical steel pivot, (the point of a small needle,) rising from the centre of a circular hole in the wooden compass-box. The bottom of this circular hole is covered with a thin film of silver or zinc, upon which a meridian line is drawn as a *diameter*. There is a round hole in the centre of the metallic circle, large enough to allow the needle to traverse freely above it, but small enough to prevent the lower part of the copper hemispherical cup from rising above the pivot of support, so as to endanger the "unshipping of the needle." The circu-

* The sea coast of China generally runs in a north and south direction, and the monsoons prevail on the coast. The south end of a Chinese compass needle is generally coloured red.

lar hole and needle are covered by glass, held in its place by a circular wooden ring. The south end of the needle is coloured red. The compass-box is of boxwood, four inches in diameter and three-quarters of an inch in depth; on its upper surface are drawn seven concentric circles, that nearest to the needle is divided into *eight* equal parts; circles Nos. 2, 3, and 4 have each 24 divisions, No. 5 has 72 equal parts, circle No. 6 appears to have 48 divisions, and the outer one is divided into 72 equal parts, the whole are inscribed with Chinese characters, in black or red ink. The machine is varnished and neatly finished. It is used in China by land and sea voyagers, by surveyors, architects, jugglers, &c.

Now, a compass of this kind is by no means convenient to steer by; because, if such a compass be placed in a binnacle, with its meridian line or one of its symbolical characters towards the ship's head, then all the other points or characters being painted *on the box* would necessarily remain in a *constant position* with reference to the ship's course; in fact, the *point* or character might with equal propriety be drawn upon the ship's deck, as upon a box fixed in the binnacle. It would be impossible for a European helmsman to steer a ship by a Chinese compass.

§ 4. The introduction of the mariner's compass into Europe was probably due to the Arabs, during the Holy Wars of the Crusaders on the eastern shores of the Mediterranean. The Italians, French, Norwegians, and even the English, have endeavoured to claim this most useful instrument as an invention of their own; but it appears to me, that such claims cannot be sustained. The earliest mention made of the compass in Europe is to be found in some old poetry, written by a certain Guyot de Provins, about the end of the 12th century, and preserved in M. S. in the Royal Library of France. Cardinal de Titri, a native of France, who had been engaged in the Crusades, and was appointed Bishop of Jerusalem, wrote an Oriental history, wherein he described the compass as being in familiar use *among the Saracens*, on the coast of Syria, although a novelty to himself.*

There is, in the Royal Library of Paris, an Arabian M.S., written in 1242, by Baïlac Kibdjaki, wherein the sea compass of this early period is described.† "We have to notice amongst the properties of the magnet, that the captains who navigate the Syrian seas, when the night is so dark as to conceal from view the stars which might direct their course, according to the position of the four cardinal points, take a basin full of water, which they shelter from the wind by placing it in the middle of the vessel; they then drive a needle into a wooden peg, or a corn stalk, so as to form the shape of a cross, and throw it into the basin of water prepared for the purpose, on the surface of which it floats. They afterwards take a *loadstone* of sufficient size to fill the palm of the hand, or even smaller, bring it to the surface of the water, give to their hands a rotary motion, towards the right, so that the needle turns on the water's surface;

* British Annual, 1837
† Letters of M. Humbolt, translated by Klaproth.

they next suddenly and quickly withdraw their hands, when the two points of the needle face the *north* and *south*. They have given me ocular demonstration of this process during our sea voyage from Syria to Alexandria, in the year 640 (or A.D. 1242)."

Here then, we have a clear description of the primative European compass, and how magnetism was communicated to the needle, stuck into a reed of straw, and made to float in a bowl of water. In those times, the Saracens had possession of the sea coasts ; but still the mariners of Syria and Egypt, had to manage their navigation under the government of their Mahomedan conquerors. Their manner of communicating magnetism to a needle made to float on water, "so as to form the shape of a cross," as described by the Arab in the above quotation, is worthy of notice. There is magic as well as superstition in it.

During a period of 200 years (from 1100 to 1300), the western world was convulsed with wars of no ordinary kind. From the farthest limits of the east, the Turks and Tartars had extended their conquests towards the west, overturned all the old governments and civil institutions ; and whilst the infidels were propagating their religious opinions by the sword, the Pope had established the Inquisition. Under such circumstances, need we be surprised, that arts declined, and science slumbered, and that we hear little or nothing of a machine which, however rude or mysterious, was in use for directing the course of ships ?

In consequence of the vast number of Crusaders that precipitated themselves on Palestine, all the maritime ports of any note engaged their vessels, either as transports for the pilgrims and the troops ; or else their ships were employed as traders, to supply the armies with provisions and stores. The Venetians, Genoese, and the people of Amalphi, rose in wealth and power, by their profitable employment during the wars between their Christian brethren and the Mahomedans of Western Asia. A bitter hatred, heightened by religious fanaticism, was kept up between them; and, generally speaking, it would have been unsafe for a Christian sailor to adopt and openly use a Saracen compass.

§ 5. The Holy Wars, or Crusades, terminated about the year 1291, leaving the mercantile navies of the Mediterranean to follow their commercial occupation. About the year 1302, one Flavious Giogo, a native of Amalphi, is said to have invented the mariner's compass. "Seven miles to the west of Salerno, and thirty to the south of Naples, the obscure town of Amalphi displayed the power and rewards of industry. The land, however fertile, was of narrow extent; but the sea was accessible and open. The inhabitants first assumed the office of supplying the Western world with the manufactures and productions of the East, and this useful traffic was the source of their opulence and freedom. The government was popular under the administration of a Duke, and the supremacy of the Greek Emperor. Fifty thousand citizens were numbered within the walls of Amalphi; nor was any city more abundantly provided with gold, silver, and the objects of precious luxury. *The mariners who swarmed* in her port excelled in the theory and practice of navigation and astronomy, and the discovery of the compass, which has opened the globe, is due to their ingenuity or good fortune. Their trade was extended to the coasts, or at least to the commodities of Africa, Arabia, and India."*

Here we have a free, rich, and enterprising mercantile and maritime people (and without an Inquisition), who, if they did not actually invent, were likely to greatly improve the compass, upon which the prosperity of their little territory so mainly depended. We have seen that the ancient compass of the Chinese, and that used by the Saracens, was altogether unfit for general purposes at sea. Any intelligent, shrewd captain, unshackled by authority, and not having the terrors of "the Holy Office" before his eyes, would soon hit upon a method to improve the compass. The man of Amalphi no doubt did improve the mariner's compass, by simply introducing a needle large enough to carry a card having the cardinal and other points painted on it. Such a compass would differ from the more ancient one in this all-important property, of indicating at once the direction of the ship's keel, and the bearing of all external objects.

§ 6. The compass of China, as has already been seen (§ 3), had its points painted on the box, which would turn along with the ship; the little magnetic needle being the only part about it that preserved its position, with reference to external objects in the heavens or on the ocean; but a compass such as we now use, or such as I believe was introduced by the Amalphian captain, having a card of the cardinal and intermediate points, borne up and traversing on a pivot, but held in a *permanent position* by the directive force of the magnetic needle to which the card was attached, would form an instrument of superior practical utility. The ship, compass-box, pivot, and every part of the apparatus is moveable under the needle and its attached card, which card remains in a constant position with reference to the magnetic meridian. So great and yet so trifling an improvement would secure to Flavio Giogo the honour due to original genius. The practical utility of such an instrument would force itself upon the public, and the successful application would soon secure its adoption by practical navigators, although many of the old superstitious coasters might continue to doubt its directive powers, and fear to speculate on magnetic doctrines, or even venture to use the new instrument, without free permission from the directors of their consciences.

From this time ships began to be improved in form and magnitude; the oar was laid aside for the sail; vessels were no longer fit for being beached; they required to carry provisions and water for longer passages; instead of coasting along shore, they shaped a direct course; sea charts had to be drawn; navigation began to assume something like a scientific appearance; and people became better acquainted with each other, and ascertained what they might advantageously exchange. A competition in maritime adventure arose in Europe, greatly to the advancement of geographical and hydrographical knowledge. We have already noticed, that from

---

* Decline and fall of the Roman Empire, vol. iv, p. 73.

the time of Homer to the end of the 13th century, discoveries had been few and far between; but when the magnetic needle had been so far improved and fitted to a compass that could be safely used at sea, we find Europeans making most rapid advances in all the sciences. Adventurers, instead of confining themselves to the shores of the then known world, advanced without fear into unexplored seas, in search of new countries.

In 1378, the Venetians discovered Greenland. The Normans discovered the Canary Islands in 1405. The Portuguese discovered the Madeiras in 1420, and they sailed to the Coast of Guinea in 1482. In the year 1489, the brother of Columbus brought maps and sea-charts to England; and in the year 1492 Columbus himself discovered America; and five years afterwards, Vasco de Gama, a Portuguese, sailed round the Cape of Good Hope, and entered the Indian Ocean. Here the Portuguese found a great number of ships, a well regulated trade on the coast of Arabia, Africa, and India, as well as within the Red Sea and Persian Gulf.

§ 7. When Vasco de Gama reached Melinda, he applied to the King for a pilot to conduct him to Calicut, on the coast of Malabar. He obtained as a pilot a native of Guzerat, and the Portuguese shewed this man an astrolabe,* but he paid small attention to it. They were greatly surprised to find this pilot well skilled in the use of the compass, the quadrant, and geographical charts; but the compasses in use in the Indian seas were found to be inferior to those in the Portuguese fleet. Hence we may infer, that the Indians, sailing with their periodical winds, had little need of great nicety in the construction of their compass. They were content with a very imperfect instrument, as the Chinese continue to be up to the present time;—the arts and sciences were probably on the decline;—whereas Providence had decreed, that the " barbarians" of Europe should emerge out of ignorance, explore the world and its wonders, shew its connections with the solar system, explain the phenomena of nature, and prove that the whole is the work of a bountiful Creator.

During a period of 180 years that the mariner's compass had been in use among the Christians of the 14th and 15th centuries, its character had been slowly but surely established, notwithstanding the intolerant and superstitious spirit of the times. Every thing likely to expand the faculties of the human mind, or appearing above the comprehension of the vulgar, was represented as profane or abominable, and dealt with accordingly. Men of superior abilities in their pursuits, instead of being patronized, were either actually persecuted, or else met with no encouragement in advancing the progress of useful knowledge. It was dangerous for men to meddle with doctrines or opinions of any kind, not sanctioned, received, or approved by the clergy; and this intolerant spirit extended to a much later period than I refer to; as witness the persecution of poor Galileo, who was thrown into the dungeons of the Inquisition at Rome, in the year 1633, for having ventured to assert that the earth was round, and turned daily on its own axis!

* A rude machine for making Celestial Observations,

It was under such unfavourable circumstances as these that maritime discovery, and the art of navigation and seamanship advanced, guided by the compass, and a few maps or diagrams of erroneous construction. Experience had taught seamen, that the compass was a faithful guide, that its needle pointed towards the pole star, and that the *card* which the needle preserved or held in an apparently permanent position, pointed out to them the course they ought to steer in returning from their commercial, exploratory or predatory expeditions.

§ 8. Christopher Columbus sailed from Spain,* in search of new regions, or in search of a new track to an old continent. Whilst sailing westward with the trade wind, on the 14th September, 1492, he discovered that the north point of the compass-needle no longer pointed towards the polar star. A deviation of this kind would take place but slowly, as the ships changed their geographical positions. The oscillations of the compass card on its pivot, whilst running down the "trade," would in a great measure tend to disguise the variation; the depression of the pole star in a more southerly latitude, cloudy weather, and other circumstances, might combine to prevent even a Columbus from observing the variation of the compass until its amount would banish all doubt about it. This discovery so alarmed the ship's company that they mutinied, asserting and believing, that they would never be able to return to Spain, since the compass itself began to deceive them! Columbus had the address to calm their fears, and command their services. But mark what followed. On his return to Spain, his statement, that the compass had *varied* in its direction, was not believed. The opposition to his correct views, and the mortification and persecution this great man had experienced, must have taught him the propriety, or rather the *expediency*, of being silent about magnetic variation, especially as his compasses had, in all probability, resumed their usual direction on the ship's return to Spain. Although other navigators had observed and announced the variations of their compasses, the *learned* of those times would not admit the fact; they rather chose to charge seamen with ignorance, and inaccuracy in their observations, than admit errors in the principles established by themselves.

§ 9. Pedro de Medina, of Valladolid, in his *Arte de Navigar*, published in 1545, *denies* the variations of the compass; but the concurring reports of commanders of ships on distant voyages obliged the landsmen, *in their closets*, to give up the point. Martin Cortez, in a treatise on navigation, printed at Seville before 1556, treats it as a thing completely established."† So here we see that a period of at least sixty years had elapsed, from the time of Columbus's observing and reporting the variation of the compass, *before* the truth of its *existence* was admitted.

About the year 1580, one Robert Norman, an Englishman, and a maker of "Compasses for Mariners," found, that however nicely he balanced his needles *before* he *magnetised* them, he was always

---

* See his history, and the difficulties he had to surmount before he obtained the means of undertaking his voyage.

† Encyclopaedia Britannica; article, Variation.

obliged to *counterbalance* that end which pointed to the north, by a bit of wax, or other substance, in order to keep the card in a horizontal position. Mr. Norman suspended a steel needle on its centre of gravity, and having touched it with a magnet, it dipped or pointed downwards, in the plane of the magnetic meridian, but about 72 degrees below the horizontal plane. This property is called the "magnetic dip." Mr. Norman published his discovery: experiments were made in various parts of the world, and it was ascertained that the magnetic needle remains nearly horizontal within the tropics, but that it *dips* towards the poles in both hemispheres. This property of the magnetic " dip," began to shake the confidence of seamen in the stability of their compass: it furnished data for philosophical speculation. The magnetic dip did not appear to derange the horizontal direction of the magnetic needle, nor to influence the variation of the compass, and consequently a ship's dead reckoning. It will, however, be seen in the sequel, that the magnetic dip is a very important element in the theory as well as practice of navigation.

§ 10. Observations began to be carefully made and recorded of the variations and dip of the needle. In the year 1580, the variation at London was 11¼° east, and in 1622, only 6¼°, and in twelve years later, it had decreased two degrees. These facts were made known to mariners by the publication of " *A Discourse Mathematical on the Variation of the Magnetic Needle;* by Mr. Henry Gillebrand Gresham, Professor of Astronomy." These announcements threw mariners into new perplexities; for in those days there were no published tables of amplitudes, or modes by which seamen might compute the sun's *azimuth*, and thereby find the variation of the compass at sea. Since the year 1580 up to the present time, the variation has been observed to change about 36 degrees towards the west; that is to say, it has changed its direction one-tenth part of a complete circle in Great Britain.*

Azimuth compasses were invented for finding the variation, and tables were computed and published for facilitating calculations at sea. Variation charts were drawn, and sea charts had the variation of the compass inserted on them. But navigators had frequent occasion to observe, that their observations of the variation made at sea did not agree with previously recorded observations made by others in the same localities; nor even did their own observations agree among themselves. The celebrated William Dampier, whose voyages and adventures gave an impulse to maritime enterprise, observed (Dampier noted everything worth notice,) discrepancies of this kind in his observations for the variation, making it either more or less than he knew it should be; and he says, " These things I confess, did puzzle me."† This was about the year 1680. About one hundred years later, and during the voyages of Capt. Cook, the same kind of magnetic disturbances were apparent in their observations. Mr. Wales states, that " variations observed with the ship's

---

* It appears from observations made and recorded in England, that the variation of the compass is decreasing, and also the dip, since 1819.

† See Nautical Magazine, 1837, p. 247.

head in different positions, and even in different parts of her, will
materially differ from one another, and much more will observations
observed on board different ships."

§ 11. The observations made during Cook's voyages would necessarily command attention, and excite observation. In the year 1790,
Mr. Downie, a master in the royal navy, when serving in H. M. S.
Glory, remarked,—"I am convinced that the quantity and vicinity of *iron*, in most ships, has an effect in attracting the needle;
for it is found by experience, that the needle will not always point
in the same direction, when placed in different parts of a ship: also
it is very easily found, that two ships, steering the same course by
their respective compasses, will not go exactly parallel to each other;
yet when their compasses are on board the same ship, they will
agree exactly." Whenever large fleets were assembled to sail under
convoy of ships of war, it was usual for the Commodore to intimate
by signal the course to be steered by the fleet during the night, and it
was usual to find these fleets much dispersed the following morning;
the compass courses of the ships composing the fleet having differed
considerably among themselves. It was no longer doubted that the
iron within a ship exerted an influence upon the compass, but it was
not known in what way this influence was exerted. It was then
supposed, and it is still believed by many, that iron *attracts* the
compass; that is to say, the north end of the compass needle is
attracted by the iron, and hence the term *local attraction*, applied to
the kind of magnetic disturbance under consideration.

Captain Flinders, R.N., had been employed in surveying Australia
and of course had ample opportunity of noticing and noting anomalous observations in magnetic bearings, and in observations made on
board, for the variation of the compass, in the *southern* as well as
in the *northern* hemisphere. On his return to England, his observations were communicated to the Admiralty, and their lordships
were pleased to direct a series of experiments to be made on the
compass, on board one of her Majesty's ships, at Sheerness. The
result of these experiments may be briefly stated.

1st. That the compass-bearing of a *distant object* was different in
different parts of the ship.

2nd. That the binnacle compass gave true bearings of a distant
object, when the ship's head was north or south.

3rd. That the greatest error in the bearing by compass was when
the ship's head was east or west.

Flinders concluded—and correctly—that the local attraction in the
same ship would be different in different parts of the world, and that
it would change with the magnetic dip.

Capt. Flinders died in 1814. A paper of his, which appeared in
the Philosophical Transactions of the Royal Society, upon "the
differences in the magnetic needle on board H.M.S. *Investigator*,
arising from an alteration in the direction of the ship's head," may
inform us of the author's views.

1st. He supposed an attractive power, with different bodies in a
ship, capable of affecting the compass, to be collected into something like a centre of gravity, or focal point, and that this point is

nearly in the centre of the ship where the iron, shot, &c. are deposited.

2nd. He supposed this point to be endowed with the same kind of *attraction* as the pole of the hemisphere where the ship might be. Consequently, in New Holland, the south end of the needle would be attracted by it, and the north end repelled.

3rd. That the attractive power of this point is sufficiently strong, in a ship of war, to interfere with the action of the magnetic poles of a compass placed in the binnacle.

§ 12: Captain Scoresby, who had commanded several ships in the northern whale fishery, and being an intelligent and well-informed man, directed his attention to the mariner's compass. His employment in high northern latitudes, where the magnetic *dip* and magnetic intensity are very great, furnished him with opportunities of making useful observations on magnetism. In his paper "On the anomaly in the variation of the needle," in the Philosophical Transactions for 1819,* we have the result of his observations; viz.—

1st. That all iron on board a ship has a tendency to become magnetical; the upper ends of the opposite bars being south, and the lower, north poles, in the northern hemisphere, and *vice versa*.

2nd. The combined influence of all the iron is concentrated in a focus; the principal south pole of which, being upwards in the northern Hemisphere, is situated in general near the middle of the upper deck.

3rd. This focus of attraction, which appears to be a south pole in *north dip*, attracts the north point of the compass, and produces the *deviation* in the needle.

4th. This deviation varies with the dip of the needle, the position of the compass, and the direction of the ship's head. It increases and diminishes with the *dip*, and vanishes at the magnetic equator.† It is a maximum when the ship's head is west or east, and it is proportional to the sines of the angles between the direction of the ship's head and the magnetic meridian.

5th. A compass placed on either side of the ship's deck, directly opposite to the *focus*, gives a correct indication on an *east* and *west* course, but is subject to the greatest deviation when the ship's head is north or south.

Captain Flinders and Scoresby were both practical and theoretical seamen and navigators, and were endowed with a considerable amount of philosophical and mathematical skill. They made careful observations on the action of the ship's iron on their compasses, and communicated the result of their valuable observations to the public. Had they been less *practical*, we might never have heard of their opinions of local magnetism; or, had they been *more mathematical* and *theoretical*, we might have been favoured with an hypothetical treatise on magnetism, founded on an imaginary base, and supported by mathematical formula contrived for the purpose. Investigations of this kind, although of the utmost importance in

---

* See Servington Savery's Paper on Magnetism in the London Philosophical Transactions for 1730.

† *The deviation caused by iron in a ship does not vanish as Scoresby says: the dip vanishes; but even where the dip — 0 an iron bolt laid horizontally North and South, will be found to have poles of opposite values at its ends.* W. W.

searching for those "laws of nature" that govern our planet, are generally beyond the comprehension of seamen, and tend rather to bewilder than to enlighten their minds.

§ 13. In the meantime, our ships continued to receive additional quantities of iron in their construction and equipment. *Iron knees* were substituted for wood, *iron tanks* for wooden casks, *iron ballast* for shingle, *iron bolts* for wooden tree-nails, *iron cables* for rope cables, *iron rigging* for hempen, and vessels *began to be built entirely of iron.* The consequence of all this was to render ships more difficult to be navigated, by reason of the local magnetism of the iron they contained. Attention was aroused to the subject, and Professor Barlow took it up, (§ 1) and received the countenance and support of Government in his investigations. Mr. Barlow made and recorded a great number of valuable experiments on the compass; he advanced a theory of magnetism, which was received with favour; and he proposed a plan for *correcting the deviation of the compass,* by means of an iron disc, placed near the binnacle, so as to *counteract* the effect of the greater masses of iron lying forward in the ship, and below the horizontal plane of the compass. This plan, if it did not entirely correct the local attraction in these latitudes, greatly lessened the errors that arose in the reckoning.

The failure of Mr. Barlow, in his endeavours to correct the compass, arose from his theoretical views of magnetism not being in accordance with experimental facts. He supposed, with Flinders Scoresby, and others, that there was a *central magnetic focus in a vessel,* which acted on the compass. He supposed that the magnetism of an article of iron depended upon the position of its centre, and not upon the position of its extremities, with reference to its action on the steering compass; and he did not believe, "that ever any particular action had been discovered between *two pieces of iron.*" He was not aware of the fact, that any two pieces of iron *will act upon each other as magnets,* as well as upon an artificial magnet.* After several trials of the correcting plates, in both hemispheres, they began to be disused, and are now almost entirely laid aside. These magnetical discussions and experimental trials were not followed by that public advantage to Navigation that might have been derived from them. Seamen again relapsed into indifference about their compasses. They began to think that iron, being so largely employed about ships, was really not so dangerous as their forefathers had taught them to regard it. The results have been as might have been expected. Our ships became more difficult to be navigated, and the masters less prudent and skilful in keeping their dead reckonings. The number of shipwrecks have consequently been greatly increased: many sailing vessels shaping a compass course, and running on shore with a fair wind; whilst steam-vessels in great numbers have, from errors in their compasses, run on rocks at the full speed and power of their engines, and have been of course destroyed, and many of their people drowned.

§ 14. We have now given a short history of the mariner's compass, and the reader will have noticed, that its improvement and the discoveries of its properties have been made but slowly. Seamen

* See Professor Barlow's valuable work on Magnetic Attraction. second edition, 1824.

have seldom been allowed to meddle with it, or pass an opinion upon its merit. The importance of the compass appears to have transferred it to the care of philosophers or ship-chandlers, and many a compass has been made " to sell," and not to steer by. There are a vast number of patent compasses, differing in price and in degree of utility, now in use; but seamen should bear in mind, that the compass *needle*, when saturated with magnetism, must necessarily point in the direction of the magnetic meridian, unless it be acted on by some external magnetic force within the vessel. The compass is influenced by three considerations arising from a single cause ; viz., its variation, its dip, its local attraction and repulsion by the ship and her contents.

§ 15. Before we treat of the practical application of the known principles of magnetism, and magnetic attraction and repulsion, it is proper that a short notice should be given of the magnet itself. The loadstone is an ore of iron, and contains as much as 80 or 90 per cent. of the pure metal. It is extensively disseminated over the globe, but is generally found in large masses in those rocks which geologists denominate as primitive. The property of the loadstone for attracting iron was well known to the ancients, and in several countries this property procured it the appellation of " leading stone, touch stone, stone which attracts iron, the stone of love," &c., names which it still retains.

In almost every country where the loadstone is known, it has received a name indicative of some inherent property in that mineral. We here add a list of nations, with the name of the magnet in the language of the country, and its signification.[*]

NATIONS.	NAME OF LOADSTONE	SIGNIFICATION.
English	Loadstone	Stone that carries a load or weight.
French	Aimant	The lover.
Spanish	Iman	The stone that attracts iron.
Portuguese ..	Iman, Padre de Cevar ..	Ditto.
Italian	Calamita	(?)
Greek	The iron stone	The stone that attracts iron.
Dutch	Geyl stein	The sight stone.
Danish	Magneit	From Magnus, the shepherd.
Swedish	Segel stein	Seeing stone, victorious stone, &c.
Icelandic ..	Leider stein	The leading stone.
Irish	Tarrangart	The drawer.
Welsh	Tywyssaen	The conductor.
Hungarian ..	Maynit-Ko	The love stone.
Russian	Magneit	From the Greek shepherd Magnus.
Polish	Magnit Kiamen	The loving stone.
Dalmatia ..	Gvozdetegh	The drawer of nails.
Finland	Randan-wetarga	The attractor of iron.
Chinese ..	Che Chy. Tohi-nan ..	The stone that directs or conducts.
Mandchow ..	Selel-edeben	The master of iron.
Japanese ..	{ Thru-ohy	Conducting stone
	Zi-ayakf	Stone for rubbing the needle.
Thibetan.. ..	Rdho-r-hatlen	The stone for the steel needle.
Tankin in ..	D'anamtcham	The stone which shews the south.
Siamese	Miiik	The stone which attracts iron.
Birman	Than-lvik-Kyouk	Ditto.
Malayan	Hatu-brani	The stone of enterprise.
Cingalese ..	Kandhoksgaluk	The stone which loves.
Arabic	Hadjarecbchaiyatm ..	The devil's or wizard's stone.
Persian	*Makusthes*	The magnetic stone.
Grecian	*Stone of Heraclia* ..	Attractor and repulser of iron, &c.
Sanscrit	*Thoumbaks*	The kisser.

This wonderful stone has, therefore, been eminently distinguished above every other kind of mineral, by names given to it by different nations, which at once convey to our minds a sense of some of its singular properties: thus, we find it called, the stone that carries a load, that loves, that attracts, that points out, that directs, that leads, that conducts, which shews the south, the nail drawer, the master of iron, the attractor and repulsor, the stone for the steel needle, the wizard's or devil's stone, the stone that loves, the kisser, the stone of enterprise, &c. These names were probably given to the loadstone at very remote periods of time and before the mariner's compass was invented, or before it was known that the loadstone possessed an almost unlimited power of transferring its own virtues to any number of steel bars, without being sensibly weakened in its magnetic intensity. If it had been known to what important uses magnetised steel bars could be applied,—as to navigation, to mining, and other important purposes,—how many more names might have been added, and every one of them conveying a new application of its principles. Magnets are now employed in working the electric telegraph. They assist in sending messages from one town to another, with a velocity greater than that with which sound travels. Messages are now correctly sent with greater speed than a cannon ball travels through the air. It is by the agency of artificial electricity and magnetism that man has accomplished this modern wonder. Official despatches are now correctly sent, by telegraph, over-land, through air, and even under rivers and the sea itself, with the rapidity of lightening; and we wonder what name an Arab would give to such an apparatus!

§ 16. It does not appear that any of the names, in the list we have collected, conveys any idea of the loadstone's having been applied to navigation ; and yet it is to this wonderful mineral, and its transferable magnetic properties to steel, and the practical application of it to the steering and conducting of ships (when all other resources fail us), that we owe the greater part of our knowledge of the world we inhabit, the ocean we have explored, and the intercourse we keep up with the remotest habitable regions.

It was believed that the loadstone *fed upon iron!* This was by no means an unreasonable supposition, since natural magnets actually acquire additional magnetic intensity by being kept in contact with iron. It is on this principle that loadstones are armed with soft iron, in order to increase their power.

It was seriously believed by the ancients, that if much iron were used in the construction of their ships, magnetic rocks on the sea shore might attract the vessels, and hold them firmly attached. Who has not read the wonderful adventures of Sinbad the Sailor, as detailed in the tales of the "Arabian Nights?" How would Sinbad's historian have managed an iron steam vessel ? The moderns, as well as the ancients, have ascribed wonderful physical, as well as moral effects to the magnet ; its properties have been applied by imposters in their systems of astronomy, astrology, divination, prediction of future events, divinity, law, physic, and fortune-telling.

The property of a magnet, in communicating a permanent magnetism of its own kind to hardened steel, and the directive power of
a freely suspended steel magnetic needle arranging itself in a north
and south direction, induced a belief that some mysterious agency in
the heavens held the compass-needle in the direction of the pole star.
It was afterwards considered, that magnetic rocks might abound in
the polar regions of the world, and draw the needle in that direction; and some supposed, that the earth itself contained a great
magnet in its central parts. The variation of the magnetic needle
proved that these views could not be correct; because, if the north
star itself had been a magnet, if the rocky regions towards the poles
had been formed of loadstones, or, if the earth had held a great
magnet in its central part, any of these agencies, if permanently
fixed in the heavens, or in the earth, would not have induced a
change in the direction of the compass-needle.

It is more reasonable to suppose, that magnetism, electricity,
and galvanism combine to form a mysterious agency pervading the
world; for electricity has been known to invert or destroy the
magnetism of a ship's compass, and by galvanism, needles may be
magnetised. We know comparatively but little of the internal
structure of the earth. The cuttings of the engineer, the punctures
of the miner, or the scratches of those who dig or quarry its surface,
have penetrated but a very small portion of the distance between the
surface and centre of the earth. We are, however, certain, that the
earth's mean density is greater than that of any rocks known to exist
near its surface. Our geological researches enable us to assert, that
the globe contains masses of metals, and metalliferous veins, abundantly disseminated among the stratified and crystallized rocks,
which form its external crust. There is evidence to shew, that the
central parts of the globe possess a higher temperature than its parts
near the surface; that subterranean fires exist in it; and that the
masses of matter composing our planet may be regarded as a
galvanic arrangement, its solid parts being connected or covered by
an ocean of brine and an elastic atmosphere. There are chemical
formations, as well as decompositions, constantly going on in it;
and the electrical, magnetical, or galvanical currents we witness
may result from the physical structure of our earth. If we adopt
this view, of the globe being a galvanic mass, many difficulties in our
magnetical speculations may vanish: for example, the changes in
the daily variation of the compass, and the great change that has
taken place in this variation during the last 260 years, may have
arisen from changes in the internal or external temperatures of the
earth in its various parts, as in Greenland and elsewhere.

§ 17. It is still believed by many, that iron and steel are the only
substances susceptible of magnetism; whereas every known substance is more or less susceptible of magnetic action. Mr. Barlow
found, that the brass box of a very fine compass, with which he had
been making experiments, had acquired a permanent magnetism.
Sir W. S. Harris, in his paper on the transient magnetic state of
which various substances are susceptible,* has given the following

* Philosophical Transactions, 1831.

table of the comparative magnetic inductive susceptibility of the following substances :—

Metals, &c.	Rolled Silver.	Rolled Copper.	Cast Copper.	Rolled Gold.	Cast Zinc.	Cast Tin.	Cast Lead.	Solid Mercury.	Fluid Mercury.	Cast Antimony.	Cast Bismuth.	Glass.	Marble.	Mahogany.	Water.
Comparative Magnetic Energy.	39	29	30	16	16	6.9	3.7	2	1	1.3	.2	0.35	0.37	0.37	0.37

Sir W. S. Harris found, that by condensing the metals their magnetic energy was increased, and that all substances receive or take up magnetism more rapidly than they part with it. The above conclusions were drawn from experiments made on metals subject to the action of artificial magnets vibrating within discs, or rings of the metals, included in the above table.

Professor Whewell in his Bridgewater Treatise,[*] has remarked, "When we consider the vast service which magnetism is to man, by supplying him with the mariner's compass, many persons will require no other proof of this property being introduced into the frame of the world for a worthy purpose.... Magnetism (he adds) has been discovered in modern times, to have so clear a connection with galvanism, that they may be regarded as different aspects of the same agents : all the phenomena we can produce with magnetism, we can produce with galvanism. That galvanism exists in the earth, we need no proof. Electricity, which appears to differ from galvanism in the same manner in which a fluid in motion differs from a fluid at rest, appears to be galvanism in equilibrium; and recently Mr. Fox[†] found by experiment, that metalliferous veins, as they lie in the earth, exercise a galvanic influence on each other. Something of this kind might have been expected from masses of metal in contact, if they differ in temperature, or, in other circumstances, are known to produce galvanic currents; hence we have undoubtedly streams of galvanic influence moving along the earth : but whether or not such causes as these produce the directive power of the magnetic needle, we cannot here pretend to decide. They can hardly fail to affect it."

The opinion here given is from *high authority*, and I cordially agree with it. The whole of the materials forming or stowed in a ship are susceptable of magnetism by inducting from the earth, sea, and atmosphere; the mechanical construction is such, that the whole fabric of the ship may be in a transient magnetic state : not only the iron, copper, lead, brass, and other metals; but also the *wood* forming the hull, fastened or covered as it really is with these metals and their oxides. Need we then be surprised, when we find the steering compass deviating from the true direction of the mag-

* On the Power, Wisdom, and Goodness of God, p. 113.
† Of Falmouth, in Cornwall.

C

netic meridian, or vibrating several points on each side of the course, when a vessel rolls from side to side?

§ 18. The following may be stated as magnetic axioms, or principles easily demonstrated to be true by experiment : viz.—

1st. The loadstone has two permanent poles, either of which will attract iron or steel, not rendered magnetic.

2nd. The poles of the same name or kind, in different loadstones, repel each other; thus, the north pole of one loadstone will attract the south pole of another; but the north poles will repel each other, as well as the south poles.

3rd. The loadstone communicates a permanent magnetism to hardened steel, and a transient magnetism to soft cast or wrought iron.

4th. The poles of the loadstone communicate magnetism, by touch, to steel, of an opposite kind to their own. Thus, if the north pole of a magnet touch one end of a steel bar, the end of a bar thus brought in contact will be a south pole, and the other end a north pole.

5th. Steel bars rendered magnetic by the loadstone become themselves magnets, and are capable of rendering other bars magnetic.

6th. The attraction and repulsion between magnets, whether natural or artificial, whether transient or permanent, are equal and mutual.

7th. If a magnet be cut into two or more parts, each part will be a perfect magnet, with a north and south pole. But the magnetic force of any magnet will not be so great *as the combined magnetic forces of all its parts*, after division.

8th. The magnetic attraction or repulsion exerted between two magnets, or between a compass-needle and any piece of iron, is not *impeded, diverted or lessened*, by the interposition of any substance whatever (iron excepted). If, for example, an iron gun were stowed in the bottom of a ship's hold, and a cargo of the most solid materials were stowed above it, the magnetic action of the gun upon the ship's compass in the binnacle would be precisely the same as if nothing had intervened between them.

9th. Hard steel retains magnetism longer than soft metal, and the harder it is made the better for retaining magnetism.

10th. If the north or south poles to two equal and similar magnets be kept in contact, their magnetism will ultimately be destroyed ; but if their opposite or contrary poles be kept in contact, their magnetism will be retained.

11th. If a steel bar be delicately poised on its centre of gravity, and then touched by a magnet, the bar or needle will arrange itself in the direction of the magnetic meridian, and in the direction of the *magnetic dip*. Thus, in England it would point magnetically *northward* and *downward*, about 70 degrees from the horizontal plane.

The following facts are of much importance to seamen, as they relate to that kind of magnetism which has been named " Inductive," *that is, not strictly of a permanent nature, although exerting the same kind of influence upon the mariner's compass as a permanent magnetism would exert.*

12th. The earth is magnetic, and gives direction to freely suspended magnetic needles; it has a north and a south pole, and since magnetic poles of opposite names attract each other, the north point of a compass-needle must be a south pole, because it is attracted by the north pole of the earth, and *vice versa* (§ 2).

13th. If an iron sphere, or any regular or irregular solid of soft iron be *imagined* as cut by a plane, at right angles to the magnetic meridian, but *in the same direction as the magnetic dip*, and if this plane be again cut by another imaginary plane passing through the centre of gravity of the iron, and at right angles to the dip, this last plane will separate the solid into two magnetic hemispheres, where will be found a north and south polarity and a magnetic equator. If the solid be an iron sphere, it will represent a miniature world, with its magnetic poles: and, if it be of considerable size, will control *a small pocket compass* when held near it. This magnetism is received from the earth, and will hereafter be more fully explained.

14th. If an iron bar, bolt, or plate, be suspended by a small thread, or by any other means, so that one end shall *dip* (in England) at an angle of about 60 or 70 degrees from a horizontal level, the iron, although not previously magnetised, will come to rest in the plane of the magnetic meridian; its lower and north end pointing nearly in the direction of the magnetic dip.

15th. If a piece of soft wrought or cast iron be taken, and held nearly parallel to the piece above mentioned, and if the upper end of one piece be made to approach the lower end of the other piece, an attractive force will be developed; but if the upper ends or the lower ends of the two pieces be brought near together, a repulsion will take place. These pieces, as well as all others, are magnetic by induction from the earth, and will act upon each other, as well as upon a compass-needle, as magnets or magnetised steel bars.

16th. If we place a long bolt or bar of soft iron, in a perfectly horizontal position, and at right angles to the magnetic meridian, or in an east and west direction, and then a delicate compass be placed near either end of the bar, the compass-needle will not be disturbed. But, if the further end be raised but one degree, the south point of the needle will be attracted. If the further end of the bar be lowered a little, the north point of the needle will be attracted and the south point repelled, in north magnetic latitude, and *vice versâ* in the southern magnetic hemisphere.

17th. If a small delicate and sensitive magnetic needle be allowed to settle in the direction of the magnetic meridian, and if a long straight bolt or bar of soft iron be laid in a north and south direction, with one of its ends near to the centre of the needle, and either east or west from it, the south point of the compass-needle will be attracted by the north end of the iron (in England), and the north point by the south end of the iron. If the end of the iron which is furthest from the magnetic equator, and nearest to the magnetic pole, be raised till the compass-needle returns to the true direction of the magnetic meridian, the *axis* of the iron will then be at right angles to the direction of the magnetic dip, or in the plane of its magnetic equator; and, by the application of a gunner's quadrant, or any other machine for measuring angles, the angle made by the direction of the axis of the iron, and the vertical, *will be found equal to the magnetic dip at the place of observation;* and the angle made by the iron, from a horizontal level, will be found equal to the complement of the dip.

18th. If we take a freely suspended or delicately poised compass-needle, and allow it to come to rest in the direction of the magnetic meridian, either pole of the needle will attract and be attracted by any part of a small piece of soft iron, such as a small nail; but if a larger piece of iron be used, say an iron bolt, the inductive terrestrial magnetism which the iron receives from the earth will control the compass-needle, and by its position will either attract or repel the needle. When the iron is small, the permanent magnetism of the needle controls it; but when the iron is large in quantity, its inductive magnetism will overcome the permanent magnetism of the small needle.

19th. The induced magnetism with which all articles of iron are saturated is received from ,the earth, and the polar axis or line supposed to join the two magnetic poles of any article of iron, not permanently magnetic, is *parallel to the direction of the magnetic dip.*

20th. Since iron is magnetic by induction from the earth, and polarised by position with reference to the direction of the magnetic dip, therefore, any change in the direction of a ship's head, or any alteration in her angle of inclination, whether in its direction or its amount, will be accompanied by a change in the polarity of the iron contained in the vessel, and by a change in the amount or direction of the local attraction, and its influence on the steering compasses.

§ 19. The magnetic dip is really an element of far more importance to navigation than has been imagined. The dip changes with the latitude: it is actually a measure of the magnetic intensity of the needle, and an index to the inductive magnetic polarity of the iron within a ship, as well as everywhere else. The dip, then, not only ought to be known and recorded on our charts, but ships should be furnished with means for finding the dip in long voyages.

§ 20. We have seen, that almost every substance experimented on has been found susceptible of a transient magnetic state by induction, and that the earth itself gives out magnetism of its own kind to solids separated from it. Without having recourse to delicate experiments by refined apparatus, the induced magnetism of iron is evident to the senses by the rudest machinery, or even without any apparatus at all. Pieces of wrought or cast iron act on each other, as magnets act on each other, exhibiting all the phenomena of attraction, repulsion, and magnetic conduction; and these metals may be made either to control a magnetic needle, or be controlled by it, in a variety of ways. This view of magnetism has not been entertained by philosophers, because they have never been in possession of a sufficient number of experimental facts.

We shall, however, .put our nautical readers in the way of satisfying themselves on these points, by means of materials used in *their ordinary* vocation.

The following table of dips will be found useful. They are the result of recent observations:—

A Table of the Magnetic Dip at the undermentioned Places; extracted from the best modern authorities:—

Place or Position.	Magnetic Dip of Needle.	Place or Position.	Magnetic Dip of Needle.	Place or Position	Magnetic Dip of Needle.
	° '		° '		° '
Plymouth	69.12 N.	Callao.........	6.14 S.	Lat........49. 0 N Lon........7. 0 W	67.18 N.
London	69.20 "	Guyaquil	9. 8 N.	Lat........40.15 N Lon........15. 0 W	64.32 "
Rotterdam ...	68.49 "	Rio Janeiro	12.54 S.	Lat........35. 0 N Lon........15. 0 W	61. 7 "
Paris	67.20 "	St. Catherine --	21.40 "	Lat........23.10 N Lon........20.45 W	53.26 "
Petersburgh ..	70. 5 "	Monte Video ..	34.51 "	Lat........12. 5 N Lon........26 20 W	42.45 "
Terceira	66. 6 "	Valparaiso.....	38. 3 "	Lat........1.12 N Lon........26.44 W	26.27 "
Rome	60.24 "	Conception	43.15 "	Lat........6.20 S Lon........32.40 W	15.37 "
Naples	55.55 "	Chiloe.........	53.59 "	Lat........10. 8 S Lon........34.18 W	10.50 "
Constantinople.	56.34 "	Port Desire ...	61.20 "	Lat........13.50 S Lon........32.19 W	1.50 "
Alexandria ...	43.46 "	Falkland Is. ...	53.20 "	Lat........16. 4 S Lon........36.18 W	1.25 S.
Bermuda	67.31 "	R. Santa Cruz..	55.16 "	Lat........24.55 S Lon........44. 5 W	17.55 "
Montreal	76.19 "	St. Helena	18. 1 "	Lat........27.26 S Lon........48.35 W	21. 7 "
Halifax } *	74.45 "	C. Good Hope..	52.54 "	Lat........34.53 S Lon........56.13 W	34. 3 "
New York	72.52 "	Seychelles.....	32. 5 "	Lat........40.40 S Lon........55.20 W	43.15 "
Washington ...	71.21 "	Penang	4.40 "	Lat........46. 0 S Lon........60.10 W	50.15 "
Antigua	48.96 "	Singapore	12. 1 "	Lat........51.32 S Lon........58. 7 W	52.38 "
Jamaica	47.19 "	Borneo	19.48 "		
Chagre	32.90 "	Amboyna	31. 9 "		
Panama	31.55 "	Raratonga Is. ..	36. 8 "		
Bahia	5.24 "	Macassar Is...	35.42 "		
Pernambuco...	13. 8 "	Tahiti.........	30.17 "		
Fernando Po ..	6.48 "	Raratinga.....	36. 9 "		
Ascension	1.39 "	Bass Straits ...	69. 8 "		
Cocos Islands ..	24.36 "	Swan River....	62.24 "		
Acapulco	39. 3 "	Sydney	62.49 "		
San Blas	45.23 "	Hobarton	70.40 "		
San Francisco..	62.28 "	Auckland Is. ..	73.10 "		
Port Vancouver	69.22 "	Kerguelens Is..	69.59 "		
San Diego	57. 6 "	Martin's Is. ...	14. 6 "		
Sitka	75.51 "	Bow Island ...	30.15 "		
Manilla	16.37 "	Majombo Is. ..	48.16 "		
Hong Kong....	30. 2 "	Simon's Bay ..	53. 4 "		

In high latitudes, seamen find that the upper and under sides of articles made of iron greatly affect their compasses; it is because the dip is also great, and the earth's magnetism, greater in high latitudes than near the magnetic equator. But upon the magnetic equator itself, the *polarity of iron and its local attraction do not vanish;* the polarity of the iron only coincides with the earth's polarity, but the iron will still continue to act on the compass under a new form. If it were possible to sail round the world on a great circle passing over the magnetic poles, the dipping-needle would perform a complete revolution in a vertical plane, and the transient magnetic polarity of the iron in a ship would also perform a revolution along with the dipping-needle.

Iron, when long exposed to the atmosphere, or action of water, gets covered with rust; the outer surface being converted into an *oxide of iron,* the magnetic properties of the metal undergo a change. When spindles of capstans, weather-cocks of buildings, &c., remain

* Ships running between New York, Canada, and Liverpool, pass over seas of variable dips and magnetic intensity, and such ships having compasses adjusted by permanent magnets are frequently lost,

long in a fixed latitude, and in a permanent position, the oxidation of the metals and the magnetic action of the earth communicate to the iron something like a permanent polarity; that is to say, a very considerable time must elapse before the iron will part with the magnetism it had acquired in its previous position.

Iron vessels furnish an illustration of this kind of magnetism. If an iron vessel's keel be laid down in or near to the direction of the magnetic meridian, the time she may remain on the stocks, and the processes of hammering and clenching of the materials together, will in some measure communicate a magnetism to the vessel of a semi-permanent character. For example, if the north end of an iron vessel, when on the stocks, be found to attract the south point of the compass, after the vessel is launched and moored in an east and west direction, it would be proper to moor such a vessel in a direction *opposite to that in which she was built*—and for some time—before the magnetism acquired in building would disappear. Attention is directed to this subject: those who intend to sail iron vessels should see to it.

If ships had continued to be built almost entirely of wood, their local attraction would never have been noticed; but the metals now enter largely into the formation of modern ships, and their compasses are proportionally affected. If the compass indicate a wrong course, and we steer by it, we run the ship into danger or actual destruction. It is surely, then, the duty and the interest of all those who have anything to do with ships, to acquire some knowledge of practical magnetism, and more especially of seamen, to learn the principles of their compass,—how they may guard against its errors, and shape a course with more confidence, and less risk, than those can possibly do who may continue to jog on in the old way, preferring to remain ignorant of causes that so materially influence their reckonings, or endanger their lives.

§ 21. We now proceed to give experimental proofs of our fundamental principles of practical magnetism. It is necessary to bear in mind, that all experiments made with *iron*, in order to exhibit its inductive and changeable magnetic polarity, should be made with iron of uniform quality throughout its mass. Iron that has been re-manufactured is unfit for the purpose, because it may contain pieces of old files, chisels, and fragments of *old steel*, which might retain magnetism in a permanent form. We should therefore select for our experiments *new iron*, that has been derived from the *ores*, and manufactured by a uniform process by machinery; that is to say, drawn out or rolled, so as to be of uniform density, and of regular form of bolt, bar, or sphere, &c.

Experiment 1*

* Sometime before the publication of these papers in the Nautical Magazine, in 1843, and after a conversation I had with Admiral Beaufort, whom I always found ready to listen to scientific or professional subjects, Sir Francis, being at the head of the Compass Committee, invited Colonel Sabine and Captain Johnson to meet me in the Hydrographic Office at the Admiralty. We met accordingly, and in order to refute certain principles generally received, and propounded by Barlow, Airy, and others, I made the experiments here detailed, and gave experimental proof, that one piece of iron will act magnetically upon another of the same kind. These principles were, however, at that time, unfavourably received by the committee.

Fig. 1.

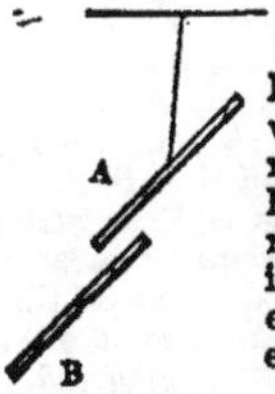

Take a piece of bolt or bar iron, a foot or more in length, and let it be suspended by a small thread, with an inclination nearly equal to the angle of the magnetic *dip*. When the twist of the thread (if it have any) has been got rid of, the bar will come to rest in the direction of the magnetic meridian, with its lower end A towards the nearest pole of the earth, and its upper end towards the magnetic equator.

Fig. 2.

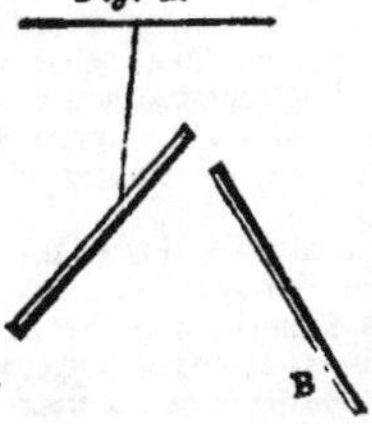

Take now a second piece of iron, B (*fig.* 1,) and hold it nearly parallel to the suspended piece A, with the upper end of B near to the lower end of A, and it will be seen that an attractive magnetic force will be exerted. But if the upper end of B be brought near to the upper end of A, (as in *fig.* 2,) a magnetic repulsion will take place.

In this experiment, it is evident that the magnetic action of the two pieces of iron upon *each other*, or upon a compass-needle, is in every way similar to the action that takes place between two natural or artificial magnets. The magnetism of the *iron* is due to its *position* with reference to the direction of the magnetic *dip* of the place, and it will be found that, by changing the position of the iron, by turning its ends, the polarity of the iron will be changed at once.

Experiment 2.

Suspend a bar of soft iron, with an inclination equal, or nearly equal to the angle of the magnetic dip, and allow it to come to rest,

Fig. 3.

(*fig.* 3,) then if a cast iron shot or shell be brought near to either end of the bar, an attraction will be manifest; that is to say, whether we bring the upper hemisphere of the shot near to the lower end of the bar, or the *lower* hemisphere of the shot near to the upper end of the bar, an attraction will take place between them because, here we present poles of opposite kinds; but it is not practicable, in this experiment, *conveniently* to exhibit a repulsion. A repulsion may, however be shewn, by bringing the shot near to the upper half of the iron; but this requires some experience in the operator.

Experiment 3.

If we take a large shot or shell of new metal, free from rust, and

tie a coloured thread round it, so as to represent a great circle—in
fact, so as to divide it into two hemispheres, as
A B, *(fig. 4,)* then let another thread, C D,
circumscribe the shot in a direction at right
angles to A B, so as again to divide the ball
into two hemispheres ; then place the sphere
upon the horizontal line S N, with the plane
of the circumscribing thread C D vertically
and at right angles to the magnetic meridian
N S ; let the shot be rolled backward till C D
becomes parallel to the *dip,* and it will then
be found, that the circumscribing thread, A B,
has divided the cast iron bolt into two magnetic hemispheres,
where C is a north and D a south pole. If now a small and
delicate magnetic needle be applied, it will be found that every
part of the ball *below* the thread A B will attract the south point
of the needle, and repel the north point; and all the upper hemisphere will attract the north point of the needle, and repel the
south point. .

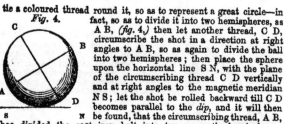
Fig. 4.

Here we have an exemplification of the local attraction of the iron
in a vessel. The magnetism developed by the shot, and which it
receives from the earth, is precisely of the same kind as that of the
gun, a tank, knee, or pig of ballast. The line C D corresponds to the
magnetic dip of the place and indicates the position of the magnetic
poles of the iron and the shot; and in fact every other thing has a
magnetism in it, and there is no such thing as *magnetism* without
magnetic *attraction* and magnetic *repulsion.* Magnetism is com-
pounded of *attraction* and *repulsion.* The intelligent mariner will
now begin to see how it is that, in these latitudes, the north point
of his compass is drawn *forward* by the iron in the vessel: he will
observe that the upper *part* attracts the north *point;* and because
all the iron is generally before the compass, and also below it, the
north point of the compass-card, which is a south magnetic pole,
must necessarily be drawn forward, so long as the *nearest* parts of
the iron in the vessel continue to retain the same kind of magnetism that the northern parts of the globe retain.

Experiment 4.

The magnetic dip being of much importance, we now proceed to
shew how it may be found, without that expensive and not over
correct instrument, the *dipping needle.*

§ 22. Take a small delicate magnetic needle, screened from the
action of the wind—a good pocket compass answers very well—and
place it on the ground, or on a table where there is no iron to affect
it. When the needle has settled in the direction of the magnetic
meridian, a rod of *pure iron,* about 8 feet in length, ¾ inch in
diameter, and perfectly straight, may then be laid in a north and
south direction, with one end within a couple of inches of the
compass, and at the same height as its pivot; the rod will be found
to derange the needle. Let C be a small pocket compass placed
on the north and south line S N, and let R be an iron rod provided

Fig. 5.

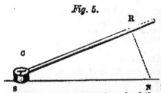

for the purpose of finding the dip of the needle; the rod R, when laid in the direction of the line S N, its end at S will attract the north point of the compass-needle; but if the north end of the rod be raised as in the figure, the compass-needle will return to its *former position;* and if the north end of the rod be raised still higher, the *south* point of the compass-needle will approach it. Let, therefore, the north end of the rod be raised so that its south end will neither attract nor repel the compass-needle, nor disturb it from pointing in the direction of the magnetic meridian,—the *axis* of the rod will then be in the plane of its magnetic equator, and consequently the magnetic dip will be at right angles to it; that is to say, the angle N S R is equal to the complement of the dip, and the angle R N S is equal to the dip itself at the place of observation. These angles are easily measured by common apparatus, *without a spirit level,* or even without the line S N being a horizontal level. A sector, or a protracter and plumb line applied to the rod R, will shew at once the angle it makes, with a vertical line. A gunner's quadrant would at once measure the angle of the dip. The dip then may be found by the ordinary means possessed by seamen; namely, by a compass, a bar of iron, and a plumb line. The principle is sound, and the application to *any extent of accuracy*, may be readily contrived by instrument makers who know their business.

Experiment 5.

§ 23. In order to shew how iron may, by its inductive magnetic property either control or be controlled by a steering-compass let a steering-compass be placed anywhere out of the influence of masses of iron. Take any number of pieces of iron of different sizes, from a small nail up to a large spike or bolt. Now, according to Experiments 1, 2, and 3, these nails are each inductively magnetic from the earth by position. Take a *small nail*, and hold it vertically near the north point of the compass, the lower end of the nail being at the same height as the compass-needle;—the nail, instead of repelling, will attract the north point, because the needle converts the nail, for the time being, into an inductive magnet, and controls the earth's inductive magnetism. Change the small nail for a larger, and as you increase the size, you will at last find a certain sized piece of iron that will *control the compass-needle*, by the induced magnetism received from the earth. Hence we infer, that although each and every piece of soft malleable iron is magnetic by induction from the earth, yet the quantity of magnetism which the earth imparts to a *small nail* may be cancelled and controlled by a magnetic compass-needle; although the natural quantity of magnetism which the earth may communicate to a *larger portion of iron* may control the compass-needle, and either attract or repel it, according to its position with reference to the direction of the magnetic dip, and the poles of the needle.

§ 24. If an iron rod or bar be placed in an east and west direction from the centre of a compass-needle, and in a horizontal position, it will not disturb the compass, nor will the needle be affected by it; but a rod or wire so placed will *conduct* inductive magnetism to the compass.

Experiment 6.

Let C (*fig.* 6,) be a magnetic needle, mounted on a pivot in the box B, and let W E be an iron rod laid in an east and west

Fig. 6.

direction, and close to the box B; the needle C will not be disturbed by the iron rod W E; but if another piece of iron, V, be held in the direction of the magnetic dip, and brought into contact with the further end of the horizontal iron W E, its magnetism will be *conducted* to the compass-needle C, and cause it to deviate from its correct magnetic meridian, (even if W E be several feet in length). If the piece V be removed a quarter of an inch from W E, and again be brought in contact, the needle will *oscillate;* thereby proving that the iron rod W E is a better conductor of magnetism than atmospheric air.

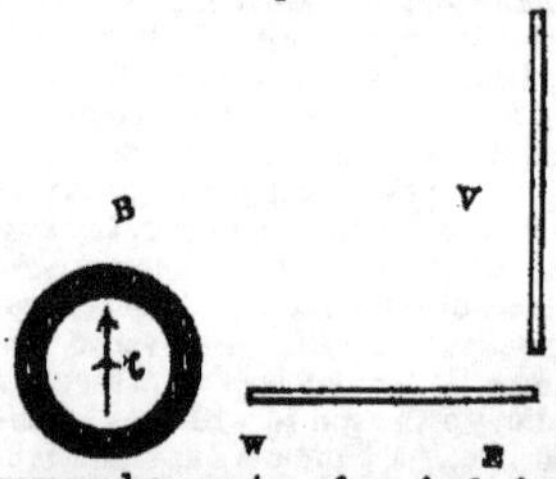

This is an important fact, because great magnetic energy may arise, and influence a steering compass, from the arrangement of the iron in a vessel. If, for example, a merchant vessel have a cargo of iron in her hold, or even iron tanks, steam boilers, or cylinders, so stowed in the hold, as to be in *contact* with an iron knee, or iron truss, bolted to the ship's side, and running upwards to the upper deck beams, such a piece of iron, being in contact with large masses of metal in the hold, would *conduct or transfer* the magnetism from below, and certainly derange the magnetic needle, and cause the compass to indicate a wrong course.

It is on this principle of magnetic conduction that separate pieces of iron, when brought into actual contact, act magnetically as a single mass. The water tanks in a ship-of-war, if stowed in actual *contact*, will act on the compass as if a single tank, of the same size as the aggregate number of small ones in the hold, occupied their places. But if the tanks be kept separate by thin slices of board, then each separate tank, &c., will retain its natural quantity of inductive magnetism, and the place of its *poles* will change with the motion of the ship.

Everybody who has been much at sea, or who has been in the habit of watching the motion of the mariner's compass, must have observed that the compass-card does not remain very steady in its bowl during bad weather. When the ship *lurches* heavily, or rolls from side to side, the compass-card oscillates several points from *the actual* direction of the ship's keel. When a ship is running before the wind, in a high sea, and rolling, perhaps, 15° or 20° on each side of the perpendicular, her compass-card may swim or vibrate a *couple of points* on each side of the course.

To remedy this oscillation of the compass-card, weight is added in the shape of wax, brass bars, &c.; for it has been considered that this vibration arises from mechanical action.*

§ 25. We have already explained how the poles of any piece of iron may be found, by means of the magnetic dip (§ 22). Now the magnetic dip has reference to the earth, and not to a ship and the iron she may contain. We may, for our present purpose, regard the earth as a fixture; but a ship when afloat and at sea is a moveable body, changing her position and direction, inclining by the force of the wind on her sails, or rolling and pitching about by the action of the waves on her hull. Now, the magnetic dip of the needle, and the consequent magnetic polarity of the iron that a vessel may contain, is always referable to a plumb line, because we measure the dip from a level; consequently the ship and her contents are constantly changing their relative positions to the dip, and also to the magnetic attraction and repulsion which every article of iron that the vessel may contain receives by induction from the earth. Whenever a ship changes her position, or her inclination, a new magnetic force is brought to bear upon the compass; and when the ship rolls alternately from side to side, equal and opposite magnetic forces act upon the compass-needle, and cause it to oscillate on each side of the true magnetic direction of the ship's keel. I beg to call the mariner's special attention to this part of our subject, and to refer him to *fig.* 6, in our last experiment, where it is shewn, that an iron bolt laid in an east and west direction, by the side of a compass (as W E), will not affect a compass-needle, even if it be within an inch or two

Fig. 7.

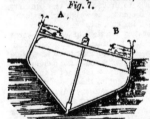

of the compass. Now, let us suppose that there is a long iron gun on each side of a ship's quarter-deck, and exactly abreast of the binnacle. When the ship's head is either north or south, the guns will be in an east and west direction, and like W E, (in *fig.* 6,) will not derange the compass-needle *so long as the ship remains quite upright.* But let the vessel be inclined as in *fig.* 7, the magnetic polarity of the two guns, and also of every bolt, bar, or nail that may be fastened through the sides of the vessel, will have changed places in the iron. Let the direction of the ship's head be north, and her inclination be to starboard, then the breech of the lee gun B will attract the north point of the needle, and its muzzle will attract the south point.

---

* See the specification of my Patent Deviation Compass, in the Appendix (A), first Edit., where it is demonstrated that in order to ensure horizontality to a compass-needle a weight is required to counteract the needles dip and therefore one part of the needle is made heavier than the opposite part.

On the other side of the ship, the breech of the weather gun A will attract the south point of the needle, and repel the north point of the compass; so that the north point of the compass-card will be drawn to leeward by the gun B, and driven to leeward by the gun A, whilst the south point of the needle, is drawn to windward by the gun A, and driven to windward by the gun B. If the water be smooth, and the ship's inclination be permanent, this kind of local attraction will permanently derange the ship's course; but if she roll from side to side, the compass-card will also vibrate on each side of the course.

If a ship's head be north, as before, but her inclination to port, by a strong easterly wind, the polarity of the guns will be inverted;

Fig. 8.

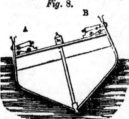

that is to say, the lee gun A (*fig.* 8,) will attract the north point of the compass-needle, and draw it towards the lee side of the ship; and the weather gun B will repel the north point, and attract the south end of the compass-needle. In this way the compass, instead of indicating a course at north, may shew a N. ¼ E. or a N. by E. course in smooth water, and in north magnetic dip; but should the vessel be running before the wind, and rolling heavily from side to side, so that, at every roll of the vessel, the inductive polarity of the iron within her be actually transferred from one side to the other, the compass-card must necessarily go on in an endless oscillation, unless means be devised to prevent it. The ordinary means resorted to by seamen is to increase the weight of the card, that is to say, to use a more sluggish instrument.

§ 26. It was owing to the vibratory motion of compass-cards mounted in wooden bowls, that copper or brass bowls were substituted for wooden ones. The fact is, that copper or brass is capable of receiving an inductive magnetism from a magnetic needle near it (§ 17); and although a brass-bowled compass may not vibrate like a wooden-bowled compass, it is beyond a doubt, that a card mounted in a brass bowl is more sluggish in fine weather.* It is even frequently necessary to have small lines attached to what are called heavy compasses, in order that the helmsman or quarter-master may agitate the compass, and cause it to traverse in light winds and smooth water. The means, therefore, that have hitherto been adopted by seamen, as well as by makers of ships' compasses, have not been founded on sound principles. These magnetic oscillations were supposed to arise from some principle in mechanics not easily understood. Whereas the vibration of the needle arises from a change of place in the magnetic poles of the iron, and other

---

* See Table of the magnetic state of which various substances are susceptible; where it will be apparent, that copper is highly susceptible to transient magnetism, and silver still more so. Copper is to Mahogany as 80 to 1, and silver, 154 to 1; therefore a silver compass bowl would be better than one of copper (sec. 17);

things that enter into the construction of the vessel and her contents; and it will be shewn, hereafter, how these vibrations, and in fact the local attraction generally, may be rectified and got rid of.*

§ 27. In order to convince seamen that the oscillation of their compass arises principally from magnetic action, the following experiments may be made in any vessel. Place a short plank (in equilibrium) upon anything so that it may rock or roll from side to side like the rolling of a ship; place a compass upon its middle, the plank being in an east and west direction, and it will be found, that the compass-card *will not vibrate*, although the plank be moved or heeled from side to side; place now a large bolt or bar of iron on the plank, on each side of the compass, and it will be seen, that the compass-card will swing or vibrate, if motion be given to the plank.

§ 28. The experiments we have been explaining *prove* beyond any doubt, that iron is magnetic, that it has magnetic poles, and that these poles are always referable to the direction of the dip of the magnetic needle, and *do not* remain in a permanent position in the iron. Any attempt that may be made to correct the local attraction of a ship's compass, or the oscillation of the needle in stormy weather, must necessarily fail, unless the operator understands clearly the philosophy of his subject. Professor Barlow failed, because he believed that the central action of all the iron in a ship remained constant in all parts of the world ;† and he did not believe that iron was polarized, as we have shewn it to be.

"I am the more anxious to establish this point," says the author,

* In the year 1841, H.M.S. Cornwallis was fitted to bear the flag of Vice-Admiral Sir William Parker, in the East Indies. At my request, Mr. Hoffmeister, the master, kindly undertook to oblige me by noting and recording, for my information, whatever changes might take place in the east iron guns of the ship, the iron fabric of the vessel, or in a piece of soft square bar iron which I supplied him for the purpose.

Mr. Hoffmeister, was simply to make the same kind of observations that I had made when in command of H.M.S. Dispatch, on the south-east coast of Africa; namely, to apply a small compass to the breech and muzzle of the guns, when the ship inclined from side to side; or to the upper or lower end of a vertical bar of iron, as the ship passed from one hemisphere or latitude to another. The iron bar with which I supplied him he suspended in his own cabin where a small compass was at hand; and he continued to make observations whilst master of the Cornwallis, and afterwards in the Jupiter, troop ship, to the command of which Sir William Parker had appointed him.

The war with China having been terminated, the Jupiter was paid off at Plymouth, and I obtained from Mr. Hoffmeister the recorded observations he had made; namely, he found—
1st. That, in north latitude in the Atlantic, the upper end of his iron bar attracted the north point of his compass, and its lower end the south point.
2ndly. That, when the ship was inclined, the highest end of a gun, attracted the north point of the compass, and the other end the south point.
3rdly. That, when the ship rolled from side to side, or tacked, the magnetism of the guns changed from one end to the other.
4thly. That, after crossing the Equator, the magnetism of the vertical piece of iron in his cabin gradually diminished, till between the 10th and 11th degrees of south latitude, and 99° 47' west longitude, he found the magnetism of the iron to be inverted, its lower end attracting the north, and its upper end the south point of his compass-needle. Mr. H., it seems, had been rather sceptical on the subject: for he made this record in latitude 22° 21' S.—"No mistake about the attraction!"
5thly. In latitude 1 59 S. longitude 89° E. the upper end of bar attracts the south point of compass: in latitude 3° N. longitude 89° E. the upper end of bar attacts the north point of compass. When on or near the Equator, in 88 E. longitude, the attractions and repulsions were recorded as very uncertain in a calm. It was, however, in this locality, when the ship was on the starboard tack, on a northerly course, that Mr. Hoffmeister observed that the polarity of the quarter-deck guns, secured in their ports at right angles to the ship's keel, had changed from + to — in one day.

† Barlow on Magnetic Attraction. Second edit. 1843, p. 271.

"in consequence of its immediate connection with the method I have proposed for correcting the errors of a ship's compass, which has been objected to, on the ground that, according to the theory we have been controverting, the central action of all the iron on board would not remain constant under all dips, and in all parts of the world; but if the hypothesis I have advanced be correct, then the central action of any irregular mass of iron will be in the centre of attraction of its surface, whatever may be the magnetic direction; and must necessarily remain the same, while the iron and the point from which its action is estimated preserve the same relative situation, as is the case with the iron of a vessel and its compass."[*]

Now, although the iron in a ship and her steering compass do actually preserve their relative position within the ship, yet the magnetic energy of the iron and its inductive polarity do not preserve their relative position in the ship, nor to the binnacle; for we have proved, that the magnetic poles of a piece of iron in a ship are referable to the earth, and not to the ship (§ § 12 25).

§ 29. The Astronomer-Royal published, in the *United Service Journal* for June, 1840, practical rules or directions, "for correcting the compasses of iron-built vessels." Now, vessels built entirely of malleable iron will *hold* inductive magnetism, and many of its pillars will, in this country, exhibit a permanent magnetism. For, independent of that magnetism which an iron vessel may receive in the progress of building, which we have already noticed, the upright bars may conduct upwards (§ 24, *fig.* 6,) a magnetism from below, or from the boilers, &c.

The directions published by Mr. Airy, although they may be of great practical utility in any iron-built vessel that may navigate the English Channel, or, in fact, around the British Isles, where the dip does not vary much; yet the plan he has proposed and the directions he has given are not applicable for distant regions. His plan is to find the local magnetism of the vessel upon its compass, by swinging the ship in the usual way, and then to correct the local attraction of the ship and her contents, by means of permanent magnets, placed at a convenient distance from the compass. His method therefore is, to correct the inductive magnetism of the malleable or cast iron fabric of the vessel, by means of permanently magnetic steel bars: that is, to correct or cancel in one hemisphere, *by a constant quantity*, a magnetic agency that may vanish, or from being positive will become negative, in the other hemisphere.[†]

---

[*] Barlow on Magnetic Attraction, p. 189.

[†] The Astronomer-Royal took offence at the concluding sentence of this paragraph, and wrote rather an angry letter on the subject. His letter and my rejoinder, with matters relative thereto, will be found in the appendix (B) of the first Edition. As however the statements made in the Nautical, have been proved by experience, I have omitted Mr. Airy's letter and my reply to it. W. W.

A Table of the Compass Deviations of H.M.S. Erebus, as ascertained before her departure from Chatham, (dip 69° 12' N.), and after her arrival at Hobart Town where the dip of the needle is 70° 40' S.

Ship's Head by Compass.	Errors of Compass.		Ship's Head by Compass.	Errors of Compass.	
	England Dip 69°12'N	Australia Dip 70°40'S		England Dip 69°12'N	Australia Dip 70°40'S
N.	0. 6 W.	1.10 W.	S.	0.28 W.	0.49 E.
N. by W.	1.12 „	0.24 „	S. by E.	0.19 E.	0. 1 W.
N.N.W.	2. 1 „	0.40 E.	S.S.E.	0.48 „	0.58 „
N.W. by N.	2.10 „	1.24 „	S.E. by S.	1.23 „	1.12 „
N.W.	3. 3 „	2.10 „	S.E.	1.53 „	1.35 „
N.W. by W.	3.28 „	2.56 „	S.E. by E.	2.21 „	2.35 „
W.N.W.	3.51 „	3.18 „	E.S.E.	2.50 „	3.17 „
W. by N.	4. 9 „	3.39 „	E. by S.	3.17 „	3.12 „
W.	4.19 „	4.15 „	E.	3.42 „	3.38 „
W. by S.	4.40 „	4.13 „	E. by N.	4.53 „	3.54 „
W.S.W.	4. 3 „	4.27 „	E.N.E.	3.46 „	3.30 „
S.W. by W.	3.24 „	4.39 „	N.E. by E.	3.18 „	3.21 „
S.W.	2.45 „	4. 6 „	N.E.	2.59 „	3.12 „
S.W. by S.	3. 8 „	3.36 „	N.E. by N.	2.16 „	2.50 „
S.S.W.	1.34 „	2.30 „	N.N.E.	1.39 „	2.28 „
S. by W.	0.52 „	1.39 „	N. by E.	0.49 „	2.19 „

The above figures have been extracted from the Philosophical Transactions for 1843. It is evident the deviations have changed their signs from east to west, and *vice versâ* (§ 12). We observe, too, that the points of no deviation do not exactly coincide in the two hemispheres. The points of the maximum deviation have also shifted a little. These points at Chatham were W. by S. and E. by N.; in Van Dieman's Land at E. by N. and S.W. by W. This may in some measure be accounted for, by the ship arriving *direct* from the north to the southern hemisphere; and *time* is required for iron to part with the magnetism it has taken up (§ 17); and this opinion is fully borne out by the fact that, on the *Erebus* returning to Hobart Town, from the vicinity of the south magnetic pole, her compass errors were somewhat greater than is shewn in the above table, and her points of maximum compass errors were at east and west. The *Erebus* was a wooden ship, with a few guns, cast iron shot, cast iron ballast, wrought iron tanks, chain cables, anchors, &c.; and we may safely assume, from the foregoing tabulated results, that the magnetism of the ship's metals changed its character by passing from one magnetic hemisphere to the other (§ 19), and that the induced poles of the ship's iron changed places in it.

The *Erebus* was commanded by Captain Sir James C. Ross, who was assisted by several competent and qualified officers. The ship was fitted out for the express purpose of making *magnetic observations* in the South Atlantic and Antarctic Oceans, for the Advancement of science, and, we trust, for the benefit of mariners. The ship and officers were supplied with useful instruments, of extreme delicacy and accuracy, and the results obtained and recorded will be received without that mistrust or doubt which attaches to statements made by persons of equivocal qualification or experience. As doubts, however, may still be entertained by sceptics, relative to an actual change in a ship's induced magnetism, by passing from one magnetic hemisphere to another, we shall give the results obtained in another vessel.

The *Terror*, discovery ship, accompanied the *Erebus* to the Antarctic regions;* and the following table exhibits the changes *her* magnetism underwent:—

A Table of the Compass Deviations of H.M.S. Terror, in England and in Van Diemans Land.

Ship's Head, by Compass.	Errors of Compass.		Ship's Head by Compass.	Errors of Compass.	
	England. Dip 69° 12′ N	Australia. Dip 70° 40′ S		England. Dip 69° 12′ N	Australia. Dip 70° 40′ S
N.	0.11 W.	0.42 W.	S.	0. 8 E.	0.11 E.
N. by W.	1.85 „	0.34 E.	S. by E.	0.51 „	0.52 W.
N.N.W.	2.31 „	1.21 „	S.S.E.	1.42 „	1.56 „
N.W. by N.	3. 9 „	2.21 „	S.E. by S.	2.30 „	2.38 „
N.W.	3.58 „	3.26 „	S.E.	3. 9 „	3.29 „
N.W. by W.	4.39 „	3.57 „	S.E. by E.	3.40 „	4. 0 „
W.N.W.	5. 8 „	4. 2 „	E.S.E.	4.34 „	4.43 „
W. by N.	5.55 „	4. 7 „	E. by S.	4.57 „	4.28 „
W.	5.55 „	4.37 „	E.	5.22 „	4.94 „
W. by S.	5.17 „	4.45 „	E. by N.	5.50 „	4.11 „
W.S.W.	4.39 „	4.53 „	E.N.E.	5.22 „	4. 7 „
S.W. by W.	3.50 „	5.23 „	N.E. by E.	4.27 „	3.27 „
S.W.	3. 8 „	4 94 „	N.E.	3.37 „	3. 9 „
S.W. by S.	2.24 „	3.32 „	N.E. by N.	2.37 „	2.37 „
S.S.W.	1.38 „	2. 4 „	N.N.E.	1.40 „	2.11 „
S. by W.	0.55 „	1.37 „	N. by E.	0.33 „	1.26 „

The reader will now see, that the *Terror's* compass deviations exhibit the same character as the deviations of the *Erebus*; the stowage and general equipment of both ship's being nearly alike.

§ 30. In iron-built vessels, the compass must always be liable to great derangement and irregularity, and less dependence should be put upon their dead reckoning. In ships built generally of timber, the local attraction upon the compass presents nearly a uniform character, although the amount of deviation may vary in different ships, and with different cargoes. The rudder is fixed at the stern, the steering wheel is near the rudder, and upon the upper deck; consequently, the compass must be placed near the helmsman, that is, upon the upper deck, and near the after end of the vessel. The principal quantity of iron in a ship will, therefore, be before and below the compass, and the nearest inductive magnetic poles in the iron will act more powerfully on the compass than the more distant and opposite poles. The result is that, in our hemisphere, and in the majority of ships, the north point is drawn forward, and in the south magnetic latitude, it is the south point of the compass that is drawn forward by the ship's local attraction; and the greatest effect takes place when the ship's head is nearly east or west.

The amount of attraction or repulsion of iron upon a ship's compass will depend upon the quantity, mass, or magnitude of the metal

* The Erebus and Terror, the two ships mentioned above, were the same vessels with which Sir John Franklyn sailed on another expedition to the Arctic regions, and never returned. After many years of anxious search for these ships, it is now known that the vessels had been frozen up in the arctic ice. Some boats, skeletons, arms, instruments and other vestiges of the crews have been found by exploring parties and brought to the Admiralty.

and its distance from the compass needle. The disturbing magnetic action of the iron increases, as its distance from the compass diminishes, in the inverse duplicate ratio of the distance ; that is to say, if we place an iron bar at four feet from the compass-needle, its magnetic action will only amount to one-fourth part of what it would be at two feet, and one-sixteenth of what its force would be at one foot distant ; and so on, the force increasing in the inverse duplicate proportion of the distance. We see, then, that a very small quantity of iron, as an iron bolt in the corner of a hatchway, or skylight, if near the binnacle, may act upon the compass as powerfully as a gun would act when secured in a port at the side of the ship, or fifty tons of iron stowed in her hold.

§ 31. The derangement of a compass, by the magnetic action of masses of iron, may be ascertained at sea by its oscillation, and by its indicating different bearings of a distant object, when the ship's head is in different directions at the same anchorage. Another sign of the existence of local attraction in a ship at sea is noticed when beating to windward, say with a northerly wind, when the ship appears, by the compass, to lie within four or five points of the wind. Whereas, when beating to the southward, she may appear to be no closer to the wind than six or seven points. These anomalous appearances in the direction of the ship's head arise from the north point of the compass-card being drawn forward, on both tacks, by the local magnetism of the vessel. Whenever these symptoms appear, a compass should be placed on the forecastle of the ship, and the magnetic direction of the ship's head on both compasses noted. The one in the binnacle will have its north point drawn forward, and the compass forward will have its north point drawn aft. Hence the correct magnetic bearing, or direction of the ship's head, will be intermediate. When doubts exist in a merchant ship about the correctness of the course, the above plan, of carrying a compass forward and comparing it with another abaft, affords an excellent check against any local attraction that may arise from receiving a new cargo, or from making changes in the stowage in a vessel.

The magnetism which the iron within a ship receives from the earth, in all latitudes, will act upon the steering compass in the following manner :—

1st. In north magnetic dip, the higher or upper parts of the iron being north poles, the north point of the compass-card (which is a south magnetic pole,) will be drawn forward in the vessel, and the south point will be repelled towards the stern, and hence the compass will indicate a course further to the northward than the ship steers ; consequently, the ship will be to the southward of her reckoning.†

2nd. In south magnetic dip, the highest or upper parts of the compass needle (which is a north magnetic pole), will be drawn towards

† In many merchant vessels of the present day, there is much iron about the wheel and steering apparatus abaft the compass. In such vessels, the north point of the compass-needle may be drawn aft in north magnetic dip.

D

the ship's head, and the north point repelled towards the stern; and hence the compass will indicate a course further to the southward than the ship steers, and she will be found to the northward of her reckoning.

3rd. In north magnetic dip, and by reason of the changeable polarity of the iron in a vessel (*fig.* 7, §25), as for example in a man-of-war, the north point of the compass-card is drawn towards the lee side, and the south point is attracted towards the weather side, whenever the ship is inclined by the force of the wind on her sails; or, in fact, by any other means, as by shifting her cargo (§ 26).

4th. In south magnetic dip, and when a ship is inclined from an upright position, the south point of the compass is drawn to leeward and the north point is drawn to windward, by the induced magnetic poles of iron being transferred from end to end of a gun or bolt in a ship's side, &c. But when a ship rolls from side to side, in regular succession, the compass-card obeys the magnetic impulses of the changeable polarity in surrounding objects, and goes on in regular oscillations.

§ 32. These are generally the conditions of the local magnetism of *all* sailing wooden vessels, and of almost all *wooden-built* steam vessels, whose compasses have not been corrected by artificial means, or removed beyond the sphere of the ship's local magnetism. If the local attraction of a ship has been correctly ascertained by swinging her round, and the amount of local magnetism noted upon each of the two-and-thirty points of the compass, then these corrections may be safely applied to the courses, so long as the ship continues to be navigated in *the same amount of magnetic dip;* but if the ship's local attraction has not been correctly found, we may deduce the following practical results from the principles we have been propounding :—

1st. Almost all wooden ships will be found nearer to the magnetic equator than the dead reckoning will place them. Thus, in England, or in north magnetic latitude, ships get to the southward of their reckoning; but at the Cape of Good Hope, or coast of Brazil, they generally get to the northward of their reckoning.

2nd. With regard to the change which takes place in the polarity of the iron in a vessel as she changes her position, and which involves important considerations to the navigator, we may deduce the following general rules.

3rd. In north magnetic latitude, and when a ship is on a wind and steering a northerly course, she is liable to be to leeward of her reckoning; but when steering to the southward, she is liable to be to windward of her computed position.

4th. In south magnetic latitude, and when a ship is on a wind, and consequently inclined, she is liable to be to leeward of her reckoning when standing to the southward; but liable to be to windward when standing to the northward.

So that in any magnetic latitude, whether north or south, a man-of-war, when on a wind, and steering towards a magnetic pole, is liable to be to leeward of her reckoning; but when standing towards *the* magnetic equator, and inclined by the force of the wind on her *sails, will* get to windward of her place by dead reckoning, when all *other things* are equal (*fig.* 7 & 8).

§ 33. We must go a little further into our subject, and remind our readers, that the general mass of the metals, being in every ship *below* and *before* its steering compass, the magnetic needle is acted on most powerfully by the *nearest* magnetic pole of the iron before it; and that the greatest effect is produced when the compass-needle is nearly parallel to the ship's beam, or at right angles to the keel. If the ship be perfectly upright, and both sides perfectly alike with regard to the iron entering into her fabric, then, when her head, or rather her keel, is in the direction of the magnetic meridian, the local attraction on the compass is = 0, but when the ship's head is either magnetically east or west, the local attraction is a maximum; the vessel being supposed *perfectly upright.* But if the ship be inclined from an upright position, the induced polarity of the iron in the vessel is transferable from side to side; and the local effect upon the compasses, under these conditions, is greatest when the ship's head is on the magnetic meridian, and least when either east or west. Now, when the ship's course is either N.E., S.W., N.W., or S.E., it is evident the local magnetism of the general mass of the vessel and her contents, as ascertained when perfectly upright, may either go to cancel or combine with the magnetism arising from a change in the ship's inclination from starboard to port, *although the direction of the ship's keel may not change.* Our investigations are assuming rather a complicated appearance, but we shall endeavour to make ourselves understood.

Ex. gra. Let a ship be steering a north-east compass course from the Longships to Milford, then it is possible that the north point of her compass may be drawn forward half a point by the local magnetism of her contents, when the ship is quite upright; consequently, although the compass might indicate a north-east course, she would actually be steering a N.E. ¼ E. course. Let it now be granted, that the wind may change to north-west, and that, by the ship's inclination, the north point of the compass-card may be drawn half a point towards the lee side (*fig.* 7); the ship, if continuing to be guided in her course by the compass, would now be steering at N.E.b.E. instead of N.E. If the wind be south-east, and the ship inclined to port, and the north point of her compass be drawn half a point to leeward, this quantity, arising from the change in the magnetic polarity of the guns, knees, or bolts in the vessel, would cancel the other kind of local attraction; and, under our conditions, the compass course would be correct on the starboard tack, but *one* point in error on the larboard tack; and in this way the quantity that the north point of the compass-card might be drawn forward, by the general magnetism, might be either cancelled or doubled by the quantity that the north point might be drawn or driven to leeward, by the ship's inclination. A commander of a vessel, unacquainted with these magnetic anomalies, might, in making one passage, pronounce his compass free from error; and on another occasion might find his vessel one mile to leeward of the reckoning, for *every five* of her distance run. What would he do? Why, place his errors to the account of a strange and unaccountable current!

On the 30th of July, 1842, H. M. S. *Vanguard,* Captain Sir David

D 2

Dunn, being some 80 or 100 miles south-west from the Lizard, and being sure of their position, shaped a course for the Lizard, with a moral conviction of correctly making the lights ahead. The weather was perfectly clear, with a fresh breeze from the north-west; but the ship got a long way to leeward of her intended course. On her arrival at Plymouth, the circumstance was mentioned to the writer, who gave Sir David Dunn and the officers a *practical demonstration* of the way whereby the north point of the *Vanguard's* compasses was drawn forward as well as to leeward, on a north-east course with a north-west wind.

Those who have clear conceptions of magnetism, and who know something of the composition and resolution of forces, will have no difficulty in comprehending how these results must influence a ship's reckoning; but those who have not paid attention to the subject, would, in order to follow our reasoning, require practical illustration by means of a model. It is one thing to enunciate a principle, and another to make the reader comprehend it.

§ 34. I might easily quote recorded instances of errors arising to the reckoning of ships under my immediate charge, in order to bear out the conclusions to which we have arrived; but I deem it more fitting to refer to cases of disaster at sea, with which the public may be better informed—cases where ships have run ashore with fair winds, steering compass courses intended to lead them clear of all danger, but which really led them to destruction. If it can be shewn, as I have shewn, that, when the local magnetism of ships is generally of such a character as to cause the vessel to deviate from her intended course in *one particular direction*, the mariner is forewarned of his danger: he knows on what side of the course the ship is likely to diverge, and takes his precautions accordingly, even if he should be in a ship where no pains have been taken about the steering apparatus, and where nothing whatever has been done to determine the deviation of the needle on the different courses the ship is liable to steer.

The British and North American mail steamer, *Columbia*, sailed from Boston, on the 1st of July, 1843, for England; and shaping a course for Seal Island, on the 2nd ran directly upon it, and became a total wreck. The sea was quite smooth at the time, although the weather was hazy. All the crew and passengers were saved by the boats. A committee of the House of Commons, appointed to enquire into the cause of shipwrecks, summoned witnesses to give evidence relative to the loss of the *Columbia*, and that evidence is worthy of notice, although it does not appear that the *committee* derived much information from it.

A fine well-appointed steamer, leaving a harbour on one day, and, with fine weather and smooth water, running the next day directly on an island, for which a correct course had been shaped, offered a problem of some interest to solve: for the magnetism of the ship and the compass were involved in it.

If the case had been investigated by a committee of experienced seamen and practical navigators, they would have disentangled extraneous evidence, and arrived at the *facts* of the case, which are very instructive,

The captain stated, in his evidence before the committee on ship-wrecks, that, during the two years he had been employed in the Company's steamers, the ships were always found to be about 20 miles to the southward of their reckoning, in crossing the Bay of Fundy, between Seal Island and Boston Lighthouse, when shaping a course of E. ¼ N. or W. ¼ S.; that this constant southerly set of the ships was ascribed to an "outset," or southerly current; that the *Columbia* shaped the ordinary course of E. ¼ N. by compass, on the 1st of July, and on the 2nd ran upon Seal Island, when they expected to be, as usual, many miles to the southward of it. The conclusion they then drew was, that the outset, or southerly current had ceased to run. Similar evidence was given by other witnesses, and the committee appeared satisfied. One member asked if the *Columbia's* compasses were correct; and was informed that they were. The witness was asked how he knew the compasses were correct; and the reply was, that the *Columbia* had recently had her compasses *made* correct by a scientific person at Liverpool, who swung the ship round, and was a whole day about it, and who placed magnets near the compass. This information appeared to satisfy the committee, and may probably please those who undertake to adjust the compasses of ships, and gain a living by the process.

In reading over the evidence, it appears clear enough, that the mail steamers employed by the Company were always found to the southward of their reckoning, in crossing the Bay of Fundy; and this southerly set of the ships was considered as the result of a current setting out of the Bay. The fact is, that these steamers would be as far as compass-deviations are concerned, very similar to other steamers. Their engines, &c. would attract the north point of the compass-needle forward, and thereby cause the vessels to diverge in a more southerly direction than their compass indicated the courses steered. If these vessels' compasses had never been adjusted by magnets or other means, their deviations in a southerly direction would have continued up to the present time; or, if the *Columbia's* compass had *not been touched by the scientific person at Liverpool*; in all probability she would have cleared Seal Island, and gone on safely in crossing the Bay of Fundy, and recording in her log-book the hypothetical "southerly set." If the gentleman who adjusted the *Columbia's* compass, and made it correct by magnets, had even said to the master, "I found your compass had considerable errors, and no doubt you generally found the ship to the southward of her reckoning. I have now made the compass sensibly correct, so that for the future your ship will not be so frequently found to the southward of her intended course." An observation of this kind would have opened the officer's eyes, and the loss of the steamer might have been avoided. The true bearing of Seal Island from Boston Lighthouse is N. 74° E., distance 225 miles; and the variation of the compass is about 10 degrees westerly. Now, the course shaped by the *Columbia*, as stated by the captain, was E. ¼ N. to which if we add 10° of westerly variation, we obtain E. 15° 37′ N.; or N. 74° 23′ E., consequently the course the ship shaped *was direct for Seal Island*. That the compass was correctly adjusted, there is little doubt, and the steerage of the vessel, in light winds and smooth water, must have been good, otherwise the vessel would have cleared the fatal island.†

A deviation of less than 5° on an E. ¼ N. course, for a distance of 225 miles, would set the vessel 20 miles from her intended direction. It is therefore probable, that the *Colnmbia's* compass had originally an error of 4 or 5 degrees on an E. ¼ N. or W. ½ S. course.

It is well known—and the public have paid pretty handsomely for their knowledge of the fact—that our ships, when running up Channel with a fair wind, run ashore more frequently on the coast of France than on the English coast; that is, they get to the southward of their reckoning. The coasting steam-vessels that regularly ply between Cork, London, Plymouth, and Dublin, know from experience—without knowing why—that they must steer from the Start to St. Catherine's E. ½ S. by compass, and from St. Catherine's to Start, W.b.N. ½ N. This they regularly do, traversing the same line in perfect safety. by steering courses not given in their sailing directions, nor diametrically opposite to each other. It will be in the recollection of our readers, that the West India mail packet, *Solway*, left the harbour of Corunna on a fine evening, and about two hours after rounding the light-house, ran upon the rocks and went down along with the greater part of her passengers and ship's company. The steamer shaped a compass course to clear the Island of Cisargas, but her compass was affected by the local magnetism and machinery of the vessel, and in the same way that the Irish packets compasses are affected. The north point was probably drawn *a point forward* on a W.N.W. course; and consequently, the vessel, instead of going W.N.W., was making a W.b.N. compass course; that is she got to the southward of her intended route, ran upon sunken rocks, and in a few minutes went to the bottom.

§ 35. In the month of November, 1842, several vessels were either lost, or in great jeopardy, near Boulogne. A ship called the *Reliance*, of 1,550 tons, from India, having about 120 persons on board, and laden with a valuable cargo, ran on shore on the coast of Merlimont, when it was believed the vessel was on the English coast. This fine ship was destroyed, and about 115 persons drowned. The *Reliance*, like many others, was a long way to the *southward* of her reckoning by steering a compass course that should have led her along the English shore. A good deal has been written about the loss of this fine ship; but no sufficient cause has been assigned. We are, however, enabled to shew *how* this fine Indiaman, laden with teas and other Indian productions, should be misled by her compass, and run upon the coast of France, even with a south-west wind. The following extract is from a newspaper, dated 1st April, 1843 :—
"During the last ten days, Mr. Kent and his associates, who purchased the wreck of the *Reliance*, near Buologne, have been busily employed in their endeavours to bring the wreck to land; they have found a chronometer, several silver and plated dishes, *and a large iron tank*, 46 feet long by 8 feet deep, and 6 feet wide."

† *See* Table of magnetic dip, (sec. 19), and observe that the "dip" at Montreal is 7° greater than at Plymouth, therefore compasses adjusted in England will be in error at Seal Island.

Having made enquiries about this huge tank, I was informed by a gentleman who knew the ship, and had seen the tank, that it was placed abaft the mainmast, and before the binnacle, and probably 18 feet below the compass. Now, here was a tank of malleable iron that would exert a magnetic action upon the compass as powerful as that of a solid mass of iron of the same linear dimensions; that is, equal to 2,208 cubic feet of iron, which would weigh 468 tons. When the *Reliance* was running up Channel, she was probably steering a *compass course* of E.b.S., and the wind being on the starboard quarter, the weather binnacle would be used. Now, the ship would have an inclination to port, and all that part of the huge iron tank that lay on the larboard side of the ship would attract the *south* point of the compass, and the weather end would draw the north point forward; because, if a line had been drawn transversely through the body of the tank, instead of being in the plane of its magnetic equator, its starboard or southernmost end would have been elevated 25° or 30° above it (see *fig.* 5); and therefore the ship was steering a compass course, under such circumstances as to be under the maximum magnetic influence of this immense tank. The tea and other things stowed above the tank would not cut off, or in any way impede its action on the steering compass, whose north point must have been drawn forward very considerably; for, at the time the ship struck, they supposed the vessel off Dungeness,* eight or nine leagues further to the northward. If this huge iron tank had never been placed in the after hold of the *Reliance*, it is probable she would have arrived safely. Its presence must necessarily have produced a deviation in her compasses, and in the direction we have indicated; for a cargo of empty iron tanks, when stowed in the hold of one of Her Majesty's Naval transports, has drawn the north point of the compass 18 degrees forward on a W.N.W. course.

In the year 1810, H. M. Ships *Nymphe*, *Pallas*, and *L'Aimable* ran ashore on the south side of the Frith of Forth, when steering a course intended to lead them up to Inch Keith. How many of our men-of-war have been lost off the coast of Holland, at a time when the pilots believed the ships further to the northward! There is no current setting upon the Dutch coast; on the contrary, there is actually an off-set by reason of the discharge from the rivers of Germany. That ships navigating high northern latitudes do generally get to the southward of their calculations is beyond any doubt; and that these errors in the reckoning arise from local magnetism is equally certain.

§ 36. Three of the West India mail steam-packets have been wrecked, and three others of the same Company have run on shore, but were floated again.† Now, if these six vessels, being under the full power of their engines, and steering a course by order of their commanders, which courses ran the ships ashore instead of clearing

---

* Ships running up Channel, go with the flood tide, and carry more than 6 hours tide with them, and are therefore liable to be ahead of the reckoning.

† Medina on Turk's Island, Isis at Porto Rico, Solway near Corunna, Medina at Saba, Teviot at Owers, Tweed on Turk's Island; also a vast number of other steamers of the Royal Navy and Mercantile Marine. This was written in 1843.

adjacent dangers, it is evident, that the courses *ordered* to be steered were either wrong courses, or else the vessels' compasses were under the influence of local attraction; and as the *whole* of these vessels were to the *southward* of their computed positions in north magnetic latitudes, it is fair to infer, that their compasses were deranged by the local magnetism of the steamers; that is the north points of their compasses were drawn forward, and the south points repelled aft. It is not in the latitude alone that these errors arise, the longitude is equally affected; for let a ship be steering a south-west course by her compass, and a distance of, say 100 miles; then, if the compass be so acted on, by the ship's magnetic attraction, that its north point be drawn forward half a point the difference of latitude made will be *greater*, and the departure *less*, than if the Compass had indicated a correct course.

In 1849, a merchant vessel took in a cargo of machinery for the Mediterranean. The master, although unacquainted with magnetism or compass deviations, possessed a good share of prudence and common sense. On sailing from Ipswich, and shaping courses from one buoy to another, he found his vessel did not go in the direction he wished her to go; in fact, he always missed the buoys; and he made this note in his log, " I begin to think our cargo affects the compasses." He got, however to the North Foreland, and in steering a southerly course parallel to another vessel, he hailed her and found that both ship's compasses were alike; but in going through the Gull-stream, he made another entry in his log, " The sailing directions give the course through the 'Gull' to the Downs, S.W. ¼ W., and our course by the compass is W.b.S." This is exactly the kind of observations that seamen should make.

This vessel steered by the land along the shore as far to the westward as Beachy Head, when the wind came aft, and a course was shaped W.N.W. for the Isle of Wight. Studding sails were set for the night; but towards morning, land was reported right ahead, on the French coast. Sail was shortened, the ship hauled to the wind on the starboard tack, for the English coast, which was made in due time.

The prudent master then groped along shore to Plymouth, in order to learn what should be done. He employed an agent, who applied to the writer, who ascertained, that on easterly and westerly courses the compass was three points in error, but correct at north and south. The agent wrote to the owner of the vessel, who replied " I cannot but think that Captain ——'s tale is a very strange one, as to his compasses being out of order through the attraction of the cargo. I have lately laden several vessels with the same sort of cargo, but I have never heard of such a tale as this of Captain ——'s before." Be that as it may, we trust the prudent captain did not in any way suffer for the steps he took to prevent risk to ship and cargo.

§ 37. The quality and amount of the inductive magnetism, which the iron in a ship receives from the earth, are always of the same kind and intensity as that of the magnetic hemisphere where the *vessel* may happen to be. When the needle dips towards the south magnetic pole, the higher parts of the iron on board will have a south magnetic polarity, and will draw the south point of the compass-card forward in a vessel when steering an easterly or westerly course; in a word, the magnetic phenomena, which we have described at some length, will, in south magnetic latitude, be found of an opposite kind to that in corresponding dips in north magnetic latitudes; there ships will generally get to the northward of their dead reckoning.

The southern seas, however, are not covered with vessels as are the waters north of the equator. The relative proportions of land and water in the two hemispheres of the earth are very unequal; for it may be seen by a map of the world, that the equator marks to the southward about nine-tenths of South America, one-third of Africa, Madagascar, Australia, New Zealand, and the Polynesian Islands. If we analyze the naval statistics, and add together the whole shipping of these southern countries, it is presumed that England alone will be found to posses a *greater number of vessels in its coasting trade.*

We frequently hear of vessels, when in south magnetic latitude, getting to the northward of their dead reckoning, on the coast of Australia, near, the Cape of Good Hope, or in South America. Several valuable ships have got on shore in South Africa through errors in the reckoning; that is, by being further to the northward than was expected.†

H.M.S. *Thetis* sailed from Rio de Janeiro, on the 4th December, 1830, having on board 800,000 dollars, for England. On the evening of the 5th December, she *ran upon Cape Frio,* and was wrecked. The weather, after the *Thetis* left Rio, was rather tempestuous; it blew strong from the south-east; a course was shaped N.E.b.E. by compass, and the ship ran against the cliffs of Cape Frio, her studding-sail boom-irons striking fire from the rocky cliffs, at a time when her dead reckoning placed the vessel thirty miles from the land.

Whenever enquiries are made about the loss of ships that have run ashore, we invariably find the *currents of the ocean* are assigned as the cause. It is even probable that a current may have set the *Thetis* to the westward; but we are certain that the local magnetism of the ship would have the effect upon her steering compasses of indicating a course more easterly and less northerly than the ship was steering. We have already demonstrated, that vessels steering a north-easterly course in the English Channel invariably get to the south-eastward of their reckoning. Now, the *Thetis* being in *south magnetic latitude,* would diverge to the *north-westward* of her reckoning, and it is not necessary to have recourse to an imaginary current to account for her loss.

§ 38. Old captains of ships,‡ who have made many voyages from

---

† The author speaks from personal experience, having passed round the Cape of Good Hope sixteen times. Many merchant vessels have got ashore on the south coast of Cape Colony, and several of her Majesty's ships have grounded or been wrecked here, by steering wrong courses; namely, the Birkenhead, Rhadamanthus, &c. One large steamer was wrecked on cape Reaffe, and another narrowly escaped the same fate on the Bird Islands.

‡ We refer to wooden-built ships, under canvass

England to India, and who have had to cross the trade winds, will, on reference to their logs, find that, on one side of the equator, errors in the longitude, by dead reckoning, have accumulated, and on the other side these errors have disappeared; the ship's longitude by dead reckoning, in the end, agreeing with that by chronometer. The ordinary track that ships pursue, in their outward voyages from England to India, is to the south-west; and so long as they continue to steer towards the magnetic equator, they are, for reasons already given, liable to be to windward of their reckoning, that is, to the eastward: but when they have passed the magnetic equator, and advance towards the south magnetic pole, where the needle dips to the southward, and being on the larboard tack, they are liable to be to leeward or to the westward of their reckoning, by reason of the ship's local magnetism having changed its character on crossing the magnetic equator.

H.M.S. *Malabar*, 74 guns. Captain Sir George Sartorius, left Plymouth in June, 1842, for Rio Janeiro. On the outward passage, her longitude by dead reckoning was carried on in her log, and the errors went on increasing in north magnetic latitude, but diminished as she approached Rio. This ship returned to Plymouth in May, 1843, her longitude by dead reckoning from Rio Janeiro to the Lizard being carried on in the log, and the error amounted to 9° 36′: *i.e.* the ship was nine degrees and thirty-six miles to the westward of her reckoning. The *Malabar* was a small 74-gun ship, built of teak wood, and armed with heavy guns; she was moreover, not a stiff ship, but inclined with a fresh breeze some ten or fifteen degrees. She had to cross the south-east as well as north-east trade winds, and consequently had to sail near 4,000 miles on the starboard tack. On her way from Rio to the magnetic equator, the *Malabar*, like the *Thetis*, would get to the northward and westward of her reckoning; and when she entered into the north magnetic hemisphere, her inclination to port would cause all her lee guns to attract the north point of the compass, and the weather guns to attract the south point. In this way, and during the transit of the north-east trade wind, and probably for a distance of 2,500 miles, the ship's compass might indicate a course nearly one point more weatherly than the vessel made through the water.

Mr. Barlow, the master of the *Malabar*, and others, informed the writer, that whilst the ship was standing to the northward on the starboard tack, they found a difference of 5° between observations, made for the variation, upon the weather side of the poop deck, and similar observations made upon the lee side of the same deck at the same time; hence it resulted, that observations made for the sun's azimuth and variation of the compass were not to be depended on. We are sometimes told, that a binnacle compass will not be influenced by a ship's local attraction, if the direction of her keel coincide with the direction of the magnetic meridian. But this opinion is incorrect, and founded on the supposition, that the local magnetism of a ship may be referred to a *central point* near the *middle of* the vessel, and that it remains constant;† an opinion at

---

† Barlow on Magnetic Attraction, edition 1824, p. 307.

AND THE MARINER'S COMPASS.

variance with the doctrine we have detailed, and the demonstrations we have given. The errors that creep into a ship's reckoning no doubt frequently arise from currents. Navigators may likewise get out, in their reckoning, by making allowances for currents that do not exist; but the great source of errors in navigation consists in bad steering, and ignorance or inattention to the magnetic attraction and repulsion of the iron in a vessel, which derange the compass courses.

Practical Rules for ascertaining the Deviations of the Compass, which are caused by the Iron in Ships, were published by order of the Lords Commissioners of the Admiralty, in 1842 and 1852, where, in § 8, I find the following sentence relating to the points upon which there are no deviation: " Experience, however, has shewn, that when the points of no deviation in any vessel have been once determined in a compass, placed as above directed, and *always used in the same place*, they may be regarded as constant, provided that no extraordinary or unusual change be made in the amount or distribution of the iron in a ship." The words here quoted are also to be found in a work by Captain Johnson,†—a work containing much valuable information obtained from observations made in H. M. ships, where good observers and excellent instruments were to be found   At pp. 104-5 of the second edition of Captain Johnson's book, we find tabulated deviations, as found in three iron steam-vessels, and five wooden steam-ships of the Royal Navy, which had their deviations ascertained in England, and afterwards at distant parts of the world, where they were swung for the purpose of again ascertaining their compass errors on each point. I have examined these tables with some care, and find that, in almost every case, the points of *no* deviation, and points of *greatest* deviation, *have not remained constant*, even though the ships have not passed from one magnetic hemisphere to another. From the tables above referred to, I have extracted and tabulated the compass errors of H.M.S. *Centaur*, swung first in England, and then near the magnetic equator; H.M.S. *Geyser*, swung first in England, and afterwards at the Cape of Good Hope, where the dip is about 53° S.; and lastly, the *Acheron*, swung first in England, and secondly at New Zealand. This table will satisfactorily demonstrate, that not only do the points of maximum and minimum compass deviations change places on the compass-card, but that the deviations from *plus* become *minus* quantities, a short time after the ships pass from one magnetic latitude to another.

† Practical Illustrations of the necessity for ascertaining the Deviations of the Compass. By Captain Edward Johnson, R.N., F.R.S., Superintendent of the Compass Department of the Royal Navy, p. 91. London: Potter, Poultry, 1852.

Table shewing the changes of Compass Deviations which took place in the undermentioned steam-ships, after leaving England and being re-swung at the following places; viz., the Magnetic Equator, Cape of Good Hope, and New Zealand. The (—) denotes the course on which the deviation = 0; the asterisk (*) that on which the deviation is greatest.

Direction of Ship's Head by Standard Compass.	Centaur, S.V. Com. Fanshawe.		Geyser, S.V. Com. Brown.		Acheron, S.V. Capt. Stokes.	
	England, 1849 Dip 69° 13' N.	Fernando Po, 1850, near magnetic Equator.	England, 1847 Dip 69° 13' N	C. Good Hope, 1850, Dip 53° S.	England, 1847 Dip 69° 13' N	New Zealand, 1850, Dip 70° (?) S
N.	2.30 E	1.55 E	2. 0 W	1.30 W	3. 0 E	0.12 W
N. by E.	3.40 „	3.30 „	0. 0 „	2.10 „	4.40 „	1.19 „
N.N.E.	5.40 „	3. 0 „	7.20 E	2.30 „	6.35 „	1.20 „
N.E. by N.	6.50 „	3.30 „	4.30 „	2.15 „	5.15 „	0.11 „
N.E.	8. 0 „	4.10 „	6.50 „	2. 0 „	3.35 „	1. 1 „
N.E. by E.	*8.10 „	4.40 „	7.20 „	1.30 „	5.40 „	7.32 E
E.N.E.	7.20 „	*4.50 „	7.50 „	1.40 „	*8.50 „	2.57 „
E. by N.	7.35 „	3.30 „	9.50 „	2.35 „	8.30 „	1.57 „
E.	6.40 „	2.10 „	*9.10 „	3.30 „	8.10 „	1.54 „
E. by S.	5.40 „	1.30 „	9.10 „	*3.30 „	7. 5 „	2.62 „
E.S.E.	4.40 „	7.10 W	9.10 „	3.30 „	5.45 „	3.21 „
S.E. by E.	3.40 „	0.30 „	7.40 „	3.15 „	4.25 „	*3.55 „
S.E.	2. 0 „	1.30 „	6.30 „	3. 0 „	5. 0 „	3.51 „
S.E. by S.	1. 0 „	1.20 „	3.30 „	1.45 „	1.40 „	3.52 „
S.S.E.	0.20 W	1.30 „	4.40 „	0.30 „	1. 0 „	3.52 „
S. by E.	1.30 „	1.10 „	3. 0 „	0.25 „	7. 5 W	3.35 „
S.	2.45 „	2.30 „	1.40 „	0.30 „	2.10 „	3.10 „
S. by W.	3.50 „	(?)	0. 0 „	7.55 E	3.40 „	1.25 „
S.S.W.	5.20 „	1.30 „	1. 0 W	2.10 „	4. 5 „	1.19 „
S.W. by S.	6. 0 „	1.35 „	2.20 „	2.20 „	4.57 „	1.10 „
S.W.	6.30 „	1.45 „	3.30 „	*2.30 „	5.50 „	1.30 „
S.W. by W.	7.30 „	2.20 „	5. 0 „	2.20 „	6. 0 „	1.32 „
W.S.W.	7.40 „	2.20 „	6.30 „	2.10 „	7.20 „	7.43 W
W. by S.	*8.10 „	*3.10 „	8. 0 „	1.40 „	*7.20 „	0.49 „
W.	8. 0 „	2.30 „	9.20 „	1.10 „	7. 5 „	0.13 „
W. by N.	7.40 „	2.20 „	10.30 „	0.30 „	6.35 „	0. 4 „
W.N.W.	6.40 „	2.20 „	9.20 „	7.10 W	6.10 „	0.58 „
N.W. by W.	5.40 „	2.20 „	*11. 0 „	0.35 „	5.20 „	1. 8 „
N.W.	4.30 „	2.10 „	9.40 „	1. 0 „	4.25 „	0. 8 „
N.W. by N.	3.10 „	1.20 „	8.20 „	0 45 „	5. 3 „	0.42 „
N.N.W.	1.50 „	0. 0 „	6. 0 „	0.30 „	1.40 „	*1.32 „
N. by W.	0.40 E	1.20 E	4.40 „	1.10 „	0.20 E	0.12 „

The above table shews, that when the ships' heads are at north or south by standard compass, there are considerable errors. In the *Centaur*, the north point of her compass is attracted towards the starboard side. In the *Geyser*, her compass is attracted to port. In the *Acheron*, whilst in England, the north point of her compass was attracted to starboard, but when she had reached New Zealand, her compass, at north or south, was attracted towards the port side of the ship; and almost all the signs of compass deviations had been changed. In these ships, it is evident there was an unequal distribution of induced magnetism on opposite sides of the ships; and hence the apparent complexity and irregularity in the amount of compass deviations on the same courses in different latitudes.

The resultant points, or points of least and greatest deviation, it is evident, have varied remarkably. For example, the greatest error of the *Geyser's* compass in England was, on a N.W. b W. course, and

= 11° westerly; at the Cape, and on a W.N.W. course, it was=
0° 10' westerly.

In the *Acheron*, the greatest errors were near E.N.E. and W.S.W. in England, but at New Zealand her least errors were N.E. b E. and W.S.W.; whence we may infer, that these resultant points, instead of remaining constant, do actually change places, in some ships, when passing from one magnetic hemisphere to another of an opposite kind, but nearly of an equal amount of dip or inclination.

Captain Johnson, in his second edition, at pages 106 and 108, makes favourable mention of two pamphlets, detailing methods proposed for reconstructing a ship's compass deviation tables at sea. These methods are grounded on the assumption, that the compass courses of greatest and least errors (resultant points), remain constant everywhere; but as the tables I have referred to clearly shew, that, in iron as well as in wooden vessels, the points of greatest and of smallest compass errors may change 50 or 80 degrees upon the compass-card, the intelligent mariner will do well to resort to frequent amplitudes and azimuths for determining the errors of his compass; for, in all cases, the difference he finds between the calculated and observed compass bearing of the sun will be the quantity he should apply to correct his course at the time of observation.

The variation, properly so called, and the deviation caused by the ship's local magnetism, will always be involved together in the observations he makes, whether by amplitude or azimuth.

§ 39. Being in command of a King's store-ship (the vessel referred to in § 1), filled with new iron water tanks, and bound down Channel for Plymouth, on the 30th December, 1818, I was of opinion, that these wrought iron tanks would exert an additional influence upon the compass. A W.N.W. compass course was shaped from St. Catherine's point for the Start. There was a fresh breeze at east, with clear weather; but at daylight, instead of making the Start, it bore N.N.E. 21 miles. In this case, the ship was at least 8 leagues further to the southward than she ought to have been, by steering W.N W. I was not then aware of the fact, discovered by Professor Barlow, that an iron tank would exert a magnetic influence equal to that of a solid of iron of the same linear dimensions.

On the 26th of March, 1803, H.M.S. *Apollo* sailed from the Cove of Cork, (now Queen's Town), with a convoy of seventy sail of merchant ships; on the 2nd of April, at three in the morning, *the frigate and forty sail of her convoy* went on shore, on the coast of Portugal, at a time when they imagined themselves three degrees to the westward of their real position. The loss of these ships may be ascribed to the local attraction of the frigate upon her compass; for about thirty sail of the convoy had, during the night, wore to the N.W., and thereby escaped destruction. The *Apollo* and her convoy had been steering for several days in a south-west direction. The north point of the frigate's compass being attracted forward, would thereby indicate a more westerly course than the convoy was making and hence the melancholy loss of so many ships together.

Being assistant master-attendant of Plymouth dockyard, in the year 1833, I was sent to Pembroke to launch and equip H.M.S.

Royal William, 120, guns, for removal to Plymouth. I had orders not to put to sea without two steam-frigates, that had directions from the Admiralty to join me at Milford. They ultimately arrived, and towed the ship down Milford Haven and out to sea. Sail was then made on the three-decker, the two steamers, and the *Pantaloon.* Symptoms were seen of the wind promising to draw round from N.W. to W. and to blow fresh, and I shaped a course for Scilly, instead of for the Land's End. The large ship, being light, clean, and well-rigged, out sailed the other vessels ; but Captain Oliver, of the *Dee,* getting within hail, enquired how long I intended to steer that course ? I replied, " Till I make Scilly." The wind increased its force, and the *Royal William* her speed to 11 or 12 knots an hour ; and the two steamers and brig were left far astern. The line-of-battle ship's compasses were perfectly correct. She *made* Scilly light ; bore up. for the Longships, passing between them and the Seven Stones ; rounded the Land's End, and ran up Channel with a strong wind backing to the S.W. ;—which enabled her to sail direct into Hamoaze without stopping.

The two steamers and brig had continued their course in the track of the large ship, under full steam and a crowd of canvas ; and in a dark and rather stormy night made the breakers of Scilly under their bows ! The brig tacked ; the steamers backed and shortened sail and stood to the N.W. till daylight ; then search was made for the line-of-battle ship, which was nowhere to be found. In the course of the day, they reached Plymouth, and not finding the ship in the Sound, were alarmed for her safety. It now turned out, that the compasses of all these vessels were under the influence of local attraction. They had never been swung, nor had their compass errors been examined in any way. The compass courses they steered, in following the *Royal William, ought to have carried the steamers and brig far to the westward of Scilly.* Their compasses indicated wrong courses : the compass in the large ship was correct.

I was ordered to return to Pembroke for another line-of-battle ship, taking a passage in the steam-ship *Salamander,* commanded by Captain, now Rear-Admiral, Austin. The adventure of the *Royal William* induced us to make many observations on the local attraction of the steamer, on our passage to Pembroke.

We took the *Camel,* dockyard lighter, in tow, and, rounding the Longships, shaped a course for St. Ann's Point, with a firm conviction on my part of making St. Govan's Head. The wind shifted to the S.W. with thick weather ; but being prepared in every way for making land, at a distance of half or three-quarters of a mile, we went boldly on, keeping a sharp look out ahead. As we neared the Welch coast, we were amused as well as instructed by the careful master of the lighter astern, standing upon the bowsprit and waving his hat for us to keep more to port. At last, and without leave, *he let go the tow rope,* and " hauled up" two or three points. We went on as before, and as was expected, made St. Govan's Head instead of St. Ann's Point.

The master of the lighter knew by *his compass* (which was right), that the steamer was steering a wrong course. My object was to prove how dangerous it was to place reliance on a compass in a steam-ship, when her local attraction was unknown; and my object in recording so many instances, even at the risk of being tedious is to impress upon the minds of navigators the imperative necessity of being ever watchful and careful of their steering apparatus.

The general use of chronometers, the correctness of our logarithmic tables for practical astronomical purposes; the accuracy of our astronomical and mathematical instruments used for naval purposes have left us little to desire for all the purposes of practical navigation so long as *the state of the weather* will allow us to make astronomical observations. But the possession of good instruments, charts, and books have made navigation so easy and accurate in *clear weather* that the necessary care and attention, under ordinary circumstances, to the *helm*, *log*, *lead*, and *look-out*, have been sadly neglected by the majority of seamen. The result has been, that worse dead reckonings are now kept than were kept before. The general adoption of iron into the construction and equipment of ships, deranges the compass courses, and a record is made on the log-board, of courses that the ship never steered; that is, the log-board only shews a compass course, instead of the magnetic course, uninfluenced by the local magnetism of the ship or her cargo.

When an observation for the latitude is obtained at noon, this latitude enables the navigator to correct his longitude by dead reckoning, and also to work out his sights for the chronometer; but when a ship has been a day or two without obtaining an observation to determine her latitude, there may be a considerable error in the latitude by account. Now, if sights be obtained for determining the longitude by chronometer, and the mean time for these sights be obtained by applying to the calculation a latitude which is not the latitude of the place where the observations were made, it is evident that the longitude obtained by such a calculation must be wrong.

H.M. Ship *Challenger* was wrecked on the coast of South America in consequence of placing too much confidence in calculations of the above description. She had not obtained an observation for her latitude for two days; her latitude by account was *erroneous* to the amount of thirty-four miles, and this latitude being used for working sights obtained for her chronometer, the computation gave a longitude *one degree* to the westward of the ship's place.†

§ 40. The makers as well as the managers of mariners' compasses should thoroughly understand the elementary principles of magnetism. The compass, like chronometers and other useful machines, should be submitted to some test of its efficiency after it is made, and before it is offered for sale, or brought into use at sea. The compass, although less costly than even a common watch, is infinitely more useful than the best chronometer. A compass is not subjected to any trial or test of its accuracy of manufacture, its magnetic intensity, or the amount of its friction on the pivot upon which

† Vide sentence of Court Martial appointed to enquire into the causes of the loss of the Challenger. Sumner's method for verifying a ship's position affords a check upon the reckoning, in cases similar to the Challenger, when the dead reckoning cannot be depended on.

the card traverses, and the amount of its directive power, compared with the weight of the needle and its card. Generally, we may remark, that neither the vendor nor the purchaser of a compass knows much about these matters; the former being satisfied if he realise a good profit, and the latter being pleased if he purchase a handsome article.

The essential qualities of an efficient compass are, great directive power combined with little weight or friction on the pivot; the compass-bowl being freely slung in jimbals attached to the box, so as to preserve a horizontal position under all cases of the ship's rolling or pitching. The steel of the needle should be of pure metal, of uniform hardness throughout, and magnetised to saturation, and the magnetic intensity of a compass-needle should be preserved by all possible care. We have said, that a compass-card should be submitted to some test. Now, a very fair and efficient test of the magnetic energy of a compass-needle is to try if two similar cards will, by their mutual magnetism, support each other's weight; that is to say, if the north point of the needle of one card be applied to the south point of a similar one, and their mutual attraction be such as to support the weight of the card, the magnetic intensity of such needle may be regarded as sufficiently strong if they mutually support each other's weight, along with the cards they respectively carry. Now, the magnetic intensity or power of every compass-needle should remain permanently a constant quantity, and this can only be accomplished by each ship being supplied with a pair of artificial magnets, of sufficient power to renovate the magnetism of the compass-needles, whenever the *test* might indicate that they required re-touching. The magnetism of a single needle is, probably, best preserved by allowing it to traverse freely on its pivot, or else to be stowed in a direction parallel to that of the magnetic dip; for if it be placed, say, with its south pole (or north point), towards the south pole of the world, or *vice versa*, its magnetic intensity will decrease; or, if a compass be placed near to a large mass of iron, in such a manner that the magnetic polarity of the iron may act so as to control the magnetism of the compass, to a certain extent, by acting in a contrary way to that of terrestrial magnetism, then will the magnetism of the compass-needle be deteriorated. If, for example, the south point of a compass-needle were placed near to the upper end of a vertical iron pillar, the needle would be deprived of a portion of its magnetism; but if the north point of the card be placed near, or in contact with the upper end of the pillar (in north dip), the magnetic energy of the compass-needle would be augmented.

We have mentioned, that the magnetic poles of the same name or kind, in any two similar and equal artificial magnets, repel each other. There is a constant effort exerted between them to obtain the mastery, and so they mutually destroy their magnetism; and if two such magnets be so situated for a considerable time, their magnetism will be nearly destroyed.†

† I recently saw in a Liverpool Steamer, a steering compass-card having two parallel dipping needles where the most powerful needle had inverted the polarity of the other, and rendered the compass useless—W.W.

There was formerly a common steering-compass, made by John
Syeds, in June 1810, in the binnacle of the Plymouth Breakwater
light-vessel; it was the only compass in the vessel, and had been
twenty-eight years in her. No care had been taken of this card,
which had remained in its box for so many years, and yet it retained
a considerable amount of magnetism. If this card, instead of having
one needle only, had been fitted with two or more equal and similar
parallel needles, we venture to assert, from the principles we have
explained, that the magnetic force of these needles would, in a comparatively short period of time, have been reduced to the natural
standard of the earth's magnetism. The spare compass-cards,
carried to sea in ships, should be stowed in boxes, and their opposite
poles connected by pieces of soft iron, in the judicious manner
recommended by Professor Barlow, and practised in the Royal Navy.

§ 41. We have already shewn, by experiment, how the changeable
polarity of the inductive magnetism, in the metals within a ship,
either draws the compass-needle quietly aside from its correct magnetic bearing, in smooth water, or else causes the compass-card to
maintain a constant oscillation of a point or two, on each side of the
course, when a ship rolls heavily from side to side in stormy weather.
(§ 25). These troublesome oscillations give rise to the most serious
obstacles to good steerage. There are but few helmsmen to be
found who can steer a ship in stormy weather, when the compass-
card is swinging about with every roll or lurch of the ship. And
when,

> "With labouring throes, she rolls on either side,
> And dips her gunnels in the yawning tide;
> Her joints unhinged in palsied langour's play,
> As ice-flakes part beneath the noontide ray:
> The gale howls doleful through the blocks and shrouds,
> And big rain pours a deluge from the clouds;
> From wintry magazines that sweep the sky
> Descending globes of hail impetuous fly,
> High on the mast, with pale and livid rays,
> Amid the gloom portentous meteors blaze."†

It is under such circumstances as are described by Falconer, that
we are taught to appreciate the worth of a good helmsman, and the
value of an efficient compass. One card is exchanged for another,
and weak needles are loaded with heavy weights, in order to lessen
the oscillations; but neither brass bars, nor brass rings, wax, paper,
nor talc, can cure the evil; for as we increase the weight and friction of the card, we only make it the more sluggish and unfit for
the helmsman's use, who, instead of being guided by the compass
in the binnacle, must ever and anon keep looking *ahead* at the
clouds, the waves, or the stars; for he finds, that a sluggish compass
does *not* indicate a change in the direction of the ship's course,
until some time after that change has taken place.

§ 42. The brass-box compass having been found to be more steady
than wooden-boxed compasses, the latter have been almost entirely
laid aside, without sufficient reason; for it is admitted, that a compass-needle mounted in a wooden box is more sensitive than one in
a brass box, when all things are equal. Let us enquire how this

† Falconer's Shipwreck.

happens.* The comparative magnetic inductive susceptibility of the
metals is very considerable ; That of copper to *mahogany* as 29 : 0,37,
or as 78 : 1 nearly. Now, all magnets have a power of communicating a certain portion of magnetism to substances brought within
their spheres of action. Thus a magnetic needle enclosed in a
mahogany box would communicate a magnetism to the box ; the
north pole of the needle imparting a south polarity to the wood near
it, and the south pole giving out a north polarity to that part of the
wooden box opposite to the south pole of the magnetic needle.
There would, therefore, result a certain amount of *attraction* between
the ends of the magnetic needle, and those parts of the wooden bowl
nearest to the needle ; and as the magnetism is taken up in *less time*
than it is *parted with*, the induced magnetism of the box would tend
to retard any oscillation of the needle. The magnetism of mahogany
is very small indeed, and can only be detected by such delicate and
elegant instruments as were used by Sir W. Snow Harris ; but the
magnetic inductive susceptibility of copper or brass being about 80
times greater than that of wood, its effects become sensible and
apparent. We see, then, that a copper or brass-mounted compass is
more steady in a gale of wind, because its box is inductively magnetised from its enclosed magnetic needle ; and, therefore, although it
be really more steady in its vibrations, it is also more sluggish in
its motion than it would be if mounted in wood instead of metal.

Every ship should have a standard compass, of a superior description, fixed in some convenient part of the ship, and raised above the
ordinary level of the binnacle, in order that bearings, amplitudes, or
azimuths, may be the more conveniently taken by it. The compass
course of the binnacle should be referable to the standard compass,
and corrected accordingly, the local attraction of the *ship* on each
point being previously found on the standard compass, as the ship
swings round the horizon.

The steering apparatus, instead of being, as heretofore, consigned
to the care of the boatswain, and stowed away in his store-room,
with iron hooks and thimbles, chain cable gear, &c., and adjacent to
the carpenter's and gunner's stores, with all kinds of metals, will in
future be placed in the master's charge, who, being entrusted with
the navigation of the ship, is of course the proper officer to have the
care of the mariner's compass—the most important of all machines.

The maker of compasses aims at making money, and he makes
his needles, not of pure hard steel, but of soft iron pointed with
steel. Such needles are easiest made and most readily magnetised,
and they require more frequent repair and cleaning. Being stowed
away without care or attention, these needles soon lose their magnetic energy, and are returned from ships to their makers rusty and
unserviceable ; and are sent for repair, for which there is a *price*,
and also a *price for re-touching weak needles*. The consequence of
all this is, that the expense of the compass department is greater
than it should be.

§ 43. The generality of sea-faring men are not so well informed
about magnetism as they should be. How can they, since philoso-

* The *Chinese* avoid the use of metal in the mounting of their compasses.

phers differ in opinion about their respective theories! We have touched but lightly on these theories, as our object has been rather to teach the navigator a few of the fundamental principles of terrestrial and inductive magnetism, upon which the practical utility of the mariners' compass depends. These principles should form a part of the navigator's education: they are essential to the practice of his art.

When Mr. Norman, the compass maker, discovered, in the year 1580 (§ 9), the tendency a magnetic needle had to dip and depart from a horizontal position, when balanced by him, previous to the needle being magnetised, he found it necessary to add a weight to the south arm of the needle to restore the card to its level. He made the first dipping needle; but it never occurred to him, that, by adding a weight to the south semi-circle of his steering compass-cards, he was adding to the errors of the mariner's compass. Strange to say, the practice has been continued up to the present time. Philosophers, too, have lent their sanction to ancient practice; so that the compass has been actually made mechanically incorrect, by erroneous views being taken of the philosophy of its mechanism.

We are taught to believe, that all heavy bodies tend or gravitate in a direction perpendicularly downwards towards the earth's centre, and that all ponderous bodies, when suspended by a string, or supported on a pivot, will have their centres of gravity directly under the point of suspension; that is to say, the point of suspension and centre of gravity will be in the same vertical plane or direction.

This opinion is fallacious. It is only true of bodies that are not magnetic. A magnetic needle, when freely suspended on a point or pivot, has its *centre of gravity drawn from beneath its point of support in all magnetic latitudes*; and hence a compass-needle, with the card and cap attached to it, may be regarded in a constant state of mechanical instability, unless it be at the magnetic equator, where the needle has no tendency to dip.

When a plain unmagnetised steel needle is fitted to its cap and card, and placed upon its pivot in the compass-bowl, its centre of gravity will necessarily be directly under its point of support; because gravity is the only force acting upon it. If all its opposite parts be similar and equal, the card will be perfectly horizontal. Let the needle be magnetised, and a new invisible force is introduced; the north end of the needle dips or points downwards, and the south point of the needle is elevated; and, consequently, the centre of gravity of the card has changed its relative position with the point of suspension. The magnetism has brought about this change, without adding any weight to the needle: the north point of the needle is attracted downwards by the earth's magnetism, and the south point is repelled upwards by terrestrial magnetism. The usual means adopted to remedy this deflection of the compass-needle is, to fix sliding weights of brass to the needle, to restore the card to a level, and thus make the card *apparently* in equilibrium.

In order to make my meaning more clear, let the following figures represent sections of a common compass-needle, with its card and cap resting upon the pivot. First, when properly balanced, then

E 2

when magnetised, and finally when re-adjusted by a weight.

Fig. 1.

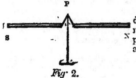

Fig. 1 is the steel needle N S devoid of magnetism : by the laws of mechanics, its centre of gravity and point of suspension in the agate P are in the same vertical.

Fig. 2.

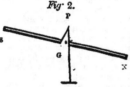

Fig. 2 is the same needle after being magnetised, and dipping towards the north. Its centre of gravity G is now drawn from under the point of suspension. It is now to the southward of a perpendicular, let fall from its point of suspension at P.

Fig. 3 is the same needle restored to its horizontal level, by means of the weight W, fastened to the needle near its south point, which weight *apparently* restores the needle to its equilibrium ; but in reality determines the centre of gravity of the mass to be *nearer to the south* than the north point of the needle. It is a self evident fact, that the south part of the card is made *heavier* than the opposite part ; and therefore their common centre of gravity is not in the same vertical with the point of suspension P, although it may appear to be so. The south arm of the needle is actually made heavier than the north arm, by the weight W.* And this is the condition in which all ships' steering compasses are made, up to the present time. The mariner's compass, then, is an imperfect instrument, in a state of constant instability, by reason of the conflicting forces of gravity and magnetism ; yet no inconvenience arises in its use, *in fine weather and smooth water.* But when the sea runs high, and the ship is rolling or plunging about among the waves, the compass, which is always fixed at a considerable height above the ships' *axis of rotation,* is carried backward and forward, and in every direction through the air at the rate of 20 or 30 feet per second. Matter, by its inertia, resists any change endeavoured to be made in its state, whether of rest or of motion. This is admitted on all sides, although doubts have been entertained about the amount or quantity of force required to put a body of a given weight in motion, or, when moving with a given velocity, to stop it. Some have maintained, that the moment of inertia, or rather the momentum of a body in motion, is as the simple velocity ; whilst others have

---

* The weight W will always be a measure of the magnetism of the needle N S, when multiplied by the leverage from the pivot P.

contended, that the force of a body in motion must be proportional to the square of the velocity with which it moves.

§ 44. It appears to me, that every effect produced must necessarily be proportional to the cause of that effect; and that, when a body is in motion, we must take into consideration the quantity of matter in the body, and the velocity with which it moves.

Now, the force arising from the quantity of matter in the body must necessarily be proportional to the quantity of matter; and the force arising from the velocity of the motion (as a cause), is necessarily proportional to the velocity of the motion. The whole force then, arising from the quantity of matter, and the velocity of its motion, must be proportional to these two causes taken together; and, therefore, in bodies of equal weight, having equal quantities of matter, and moving with equal velocity, their momenta will be equal.

If the force of a ponderous body in motion were as the *square of its velocity*, then would a cannon ball of 24 pounds, when moving with a velocity of 100 feet in a second, be arrested in its flight by coming in collision with a ball of only 6 pounds, and moving in an opposite direction with a velocity of 200 feet in a second. We know experimentally, that it would require a bullet of 12 pounds in weight and moving with a velocity of 200 feet in a second to stop it. To imagine that the motive force of a body of a given weight, when moving with a given velocity, is as the *square* of that velocity, is to imagine that the force which arises from the velocity is equal to the square of itself.

If this reasoning cannot be controverted, it is useless to suppose that any arrangement of sliding weights upon the needle of a steering compass, whether they be placed nearer to or further from the point of suspension, can be made to balance the card on a level, and at the same time make the inertia of the opposite semi-circles of the card perfectly equal under all circumstances.

If our reasoning be applied to a compass-card with an unmagnetised needle, mounted on its pivot in the usual way, in a compass-bowl, its centre of gravity would be in the same vertical as its point of suspension; all its parts would be in equilibrium by the force of gravity; and although the card might vibrate by reason of its pendulous centre of gravity when put in motion, yet there would be no tendency in the card to oscillate or swing in a horizontal plane; because the inertia of all its opposite parts would be equal, and the centre of its inertia would be in the same vertical line as the point of support; in a word, there would be no reason or cause why the card should not remain on its pivot, without turning round or oscillating upon it.

Let now the needle receive the magnetic touch in north magnetic latitude. The needle arranges itself in the direction of the magnetic meridian; the needle dips, and drives the common centre of gravity of the card to the southward of its point of suspension. There is now a greater quantity of matter to the southward than there is to the northward of the pivot's point; and if we put the compass in motion alternately in an east and west direction, the south semi-circle of the card *lags behind*, by reason of its superior weight ; and, if we bring it suddenly to rest, the south part of the card *goes on*, by reason of its momentum being greater than that of the opposite semi-circle,—the directive force of the magnetised needle striving to get the mastery of the disturbing mechanical force arising from gravity, and the unsupported position of the common centre of gravity of the card and its appurtenances.

§ 45. In high magnetic latitudes, where the dip is great, the mariner's compass, in its present construction, must be regarded as a very imperfect instrument, by reason of the disturbing forces that exist between gravity and magnetism. Gravity acts in a line perpendicularly downwards ; but magnetism acts in a line parallel to the direction of the magnetic dip, and varies in different latitudes from a horizontal direction, at the magnetic equator, to the vertical one at the magnetic poles. When a compass-needle is but slightly magnetic, and the card is comparatively heavy, such a compass will be very sluggish in smooth water, but tolerably steady in stormy weather; but if the card be a light one, and its needle a powerful magnet, then will the compass be sensitive and serviceable in smooth water and fine weather, but unsteady and of little use in a storm. The reasons are obvious enough : a weak magnetism in a heavy card but slightly deflects the centre of gravity from its point of support; whereas a powerful magnetism in a light card has the effect of deflecting the centre of gravity very considerably, and consequently, the momenta of inertia in the opposite parts of the card must be very unequal, when the ship rolls heavily in high magnetic latitudes, and is steering a course near the direction of the magnetic north or south points of the compass.

All kinds of plans have been tried to diminish this oscillation of sea compasses. Extra gimbals have been applied ; cards have been placed in rectified spirits of wine; heavy weights have been added ; compass-cards have been mounted on their pivots like toad-stools on their stalks, and supported by metal braces like a lady's *parasol*, or a gentleman's *parapluie* ; *chain cables* have actually been shackled on to the lower side of compass-needles, in order that the ends of the chains might drag about in the compass-bowl, and as it were, moor the needle in the direction of the ship's course. Of course, all these patent contrivances were of no practical utility. How could they be ? Can the physician cure the malady without knowing something of the nature of the patient's disease ?

Captain Johnson, superintendent of the compass department, states at pages 77—80 of the 2nd edition of his book, the result of a trial of different kinds of compasses in bad weather, on board the *Garland* steamer, between England and Ostend, in 1850, and made by direction of the Lords Commissioners of the Admiralty.

The following compasses were then tried in the *Garland:*—1. A standard compass on its elevated pillar. 2. Mr. Dent's large compass in the binnacle. 3. A liquid compass by Mr. Preston. 4. A compass on treble gimbals, with point of suspension elevated one quarter of an inch above the plane of the card ; its pivot passing through a brass collar *similar* to Mr. Walker's plan (but certainly not my plan). 5. A compass with an edge bar needle, ruby point, a speculum metal cap, and four ivory pins on the face of the card, their points being ⅛ of an inch from the glass cover when the card was on a level, or parallel to the cover—on Captain Johnson's plan. 6. A compass by Messrs. Grey and Kean, of Liverpool, having a card suspended in double central gimbals, with two needles on axes so as to dip.

The results obtained, as stated by Captain Johnson, who was accompanied by Mr. Dent, in the *Garland*, are as follows :—That Nos. 2 and 3 were the steadiest under all circumstances ; but that when the vessel was running before the sea, or with the sea on the quarter, Mr. Dent's compass became unsteady : that compass No. 5, on Captain Johnson's plan, and No. 4, in some measure resembling one of mine, were sufficiently steady to steer by, till the sea became turbulent and struck the ship, when the standard compass vibrated *twelve points :* Nos. 4 & 5, from one to one and a half points : No. 6 (the Liverpool dipping compass), swung half round : and ordinary compasses spun wholly round. To remedy these oscillations of the Admiralty compasses, Captain Johnson states, that he has applied four glass beads, suspended from the glass compass-bowl cover, to within ⅛ of an inch of the card.*

§ 46. We have now shown, that the mariner's compass, in its ordinary construction, is an imperfect instrument, and we have explained the nature of its imperfection ; we have also shewn in what way the induced magnetism of a ship's iron acts upon the compass. It remains to be shewn how the mariner's compass itself may be rectified, and how the magnetism of a ship's iron may be cut off from acting upon the compass, by means of the same kind of materials that disturb it in different magnetic latitudes, and not by permanent magnets.

I stated in a former part of our subject, that an efficient compass should possess a great magnetic energy, with small weight and friction on the point of suspension ; and the reader will have noticed that these conditions appear to be incompatible ; they may, however, be united in the same compass. A small magnetic needle is comparatively more powerful than a large one, consequently there is some sized needle better than any other size ; for if we go on to augment the quantity of steel, the ratio of the weight to the magnetism will increase ; and if we make the needle too small, the ratio of the directive power of the needle to the weight of the card it has to carry will be too small. Now, I have concluded, that in a steering compass, the needle, with its card and cap complete, should not weigh more than 1000 grains, nor less than 800; that of this quantity the card should be made as light as possible ; and that the magnetic energy of the needle should be of a permanent nature ;

* Here we find it officially reported, in 1850, that in a heavy sea the Committee standard compass vibrated 12 points, and that glass beads were to be suspended from the cover of the steering compass, so as to touch and retard an unsteady compass-card. Twenty years ago, the writer made an official statement, that the standard compasses at £25 each possessed the good qualities in smooth water, and the bad ones in storms, that a 13s. 6d. compass possesses. The Thunderer's and the Asia's compasses swung 8 points—the Garland's 12 ! So much for improvements in ten years.

made so by hardening the steel needle; and that a test of its magnetism should be, the *lifting* and *suspending*, by its magnetism, of a card and needle similar to itself.

The mica of the card should be sound, thin, and uniformly covered with thin paper, cemented to it, and varnished.   The cap should be jewelled and fixed perpendicular to the plane of the card.   The needle should be firmly secured under the card in the usual way; but all sliding weights and unnecessary matter dispensed with; the whole should be made in perfect equilibrium, with the plane of the card perfectly horizontal.   When the point of suspension, centre of gravity, and centre of percussion are in the same straight line, and perpendicular to the plane of the card, all these conditions must be secured *before any magnetism is communicated to the needle*.   The operation requires great care and correctness.   No part of the card must be heavier than another : it must be found *experimentally* to be in perfect equilibrium.   A *compensation regulator* is required to be fixed to the under side of the needle, directly below the point of suspension, made moveable and secured like the compensation weights on the balance of a chronometer; through the regulating compensator the spindle of the compass-card must pass, and up to the ruby in the cap, which must be centred and polished to receive the point of the vertical pivot.*

When these preliminary operations are performed, the card should be placed upon its pivot, and held in a slanting direction, to ascertain if it be in equilibrium in all kinds of positions; in fact, that all its parts be equally heavy.   When this is done by gravity only, the momenta of all its opposite parts will be found equal; and therefore there can be no tendency in the card to turn round on its pivot by any change of horizontal motion of the vessel.   The needle will also be prevented from vibrating ; and when finally magnetised, the only force that can act upon it will be the horizontal and directive magnetic force.   The needle cannot dip, because the pivot, passing through the compensation collar, prevents the dip, and keeps the common centre of gravity directly under the point of suspension, and preserves the stability of the instrument.   There cannot be any other force acting upon the needle than that of magnetism ; namely, the terrestrial and directive magnetism exerted upon the needle by the world, and the induced magnetism of the ship's iron, which has been discussed in a former part of this paper.

§ 47. We see, then, that the philosophy of the mariner's compass is mainly limited to the needle, and its mode of suspension.   There is as much difference between a common and a rectified compass, as there is between a common watch and a chronometer.   There are very few compass-makers who know anything about their business. These important and most useful of all nautical machines must henceforth be finished by a superior class of workmen.

The compensation regulator on the under side of the needle may either be fastened to the needle itself, or to the under side of the

* For further information on this subject, see specification of my compass for ascertaining and indicating the errors and deviations of the mariner's compass in Appendix A, first edition.

ordinary cap; the hole through which it passes may either be circular or square. In either case, the friction will not be greater than one 1-hundredth of the friction at the point of suspension; when the magnetic dip is 70 degrees; and when the dip vanishes, the friction in the compensator will vanish also. We have said, that the aperture in the compensation collar may either be round or square. It should be a *hair's breadth* greater than the diameter of the pivot. It will be evident to every person conversant with mechanics, that a cylindrical spindle will meet with less friction by working in a square hole, than in a circular one. A circular one is, however, of easiest construction, and more easily jewelled.

We now proceed to describe the mode of suspension of the card in the compass-boxes now in use; for any kind of compass *now in use*, may have new cards fitted to it, and be made an efficient instrument for all kinds of weather. The way of mounting a compass-card is simply to place it upon the top of a pivot which rises vertically from the the bottom of the compass-bowl,—the bowl being swung in gimbals to preserve it in a horizontal position. The card thus suspended on the agate of its cap is free to vibrate in any direction; to traverse by magnetism, or to oscillate horizontally by mechanical or magnetic action; since the centre of gravity of the mass, as has been shewn, will always be on that side of the pivot which is opposite to the end of the needle that dips.

Instead of suspending the card immediately upon the ordinary vertical spindle rising from the bottom of the compass-bowl, I place a cone of brass, with its apex made to rest upon the vertical pivot. It is jewelled and highly polished. Upon the top of the brass cone, a pivot made to receive the rectified card is firmly screwed in the direction of the axis of the cone; so that the spindle in the bottom of the compass-bowl, and that upon the top of the cone shall be in the same straight line. The card being fixed on its pivot, the weight of the cone serves as a counterpoise to the weight of the needle, and acts also as an additional pair of gimbals.

§ 48. Let us now examine the conditions of this arrangement. The cone being of brass, or other compound metal, will be devoid of permanent magnetism, and being suspended on a fine steel point, it will act in every respect as an additional pair of gimbals to the compass; being at perfect liberty to traverse on its point of support, it promotes the free action of the card above it,—acting, if I may so say, as a friction roller to the axis upon which the needle traverses. This mode of double suspension, whilst it promotes the free action of the needle to persevere in its magnetic meridianal direction, when the direction of the ships' course varies, the double suspension reduces the number of vibrations, and their angles of deflection, when the needle is diverted from its meridian. The cone is not a magnetic body, and its centre of gravity is directly under its point of suspension. Now, if a rotary motion be communicated to the cone, it will have a tendency to go on in the same direction by reason of its inertia. But if the needle, mounted upon its pivot above the cone, be deflected by a magnet, say 90 degrees from its magnetic meridian, and then set free, the cone will, to a certain extent, revolve on *its* pivot, in the same direction as the needle; but the magnetism of the needle brings it *back*, whilst the cone has a tendency to *go on*: the cone, therefore, exerts a force to retard the oscillations of the needle. I found, experimentally, that a compass-card, by single suspension, made 12 vibrations when deflected 90 degrees from its meridian; and when the cone under it was set free, the vibrations were reduced to six when deflected as before.*

A committee was appointed by the Lords of the Admiralty, for the improvement of ship's compasses. This committee introduced a new set of instruments, of a superior and costly description. Every possible care and attention was bestowed upon these compasses, and the cost of each standard compass according to the navy estimates, was £25. The cards are suspended on a ruby in the top of the cap, and some of the cards have *eight* powerful magnetic needles laid parallel to each other, and placed with their edges downwards. Here we have a single friction, a single card of a light and elegant description, with eight powerful magnets beneath it. The common centre of gravity of these needles is necessarily at a considerable distance below the point of suspension, by reason of the needles having their *edges* vertical instead of horizontal. These powerful compasses are "tremblingly alive" to all magnetic agency, and on shore are accurate and useful beyond all former comparison. At sea, in fine weather and smooth water, they have been reported as "invaluable;" but in running before the wind with a heavy sea during a *gale*, "they were found to be *useless*; and had there been none others on board, the ship must have been brought to the wind and steadied to ascertain any bearing."

Report on the standard compass of H.M.S. Thunderer. Furnished by order of her late Captain, Daniel Pring. Report on cards A, D, C of Standard compass, G 40.

1843.—February 28th. Sailed from Cork; wind northerly No. 4; steering S.W.; extreme rolling 9° each way; card D swinging one point.

March 1st. Latitude 50° 20′ N., longitude 8° 57′ W.; wind S.E. No. 5; steering S.W.; ship lurching to leeward, 11°; card D swinging 3½ points.

March 4th. Latitude 43° 44′ N., longitude 12° 2′ W.; wind variable to northward, No. 1 to 0; head to southward and westward; ship rolling 20° each way; card C swinging from S.b.W. to W.b.S., at 10° roll, 3½ points : tried card A, extreme swing 3½ points, at 10° roll one point; put on common Azimuth card; very steady, as were those in the binnacle.

March 9th. Ship rolling 5° each way; card D swung one point.

March 12th. Wind N.E. No. 4; steering S.W.; extreme rolling 8° each way; card D swinging ¾ point.

March 15th. Latitude 29° 18′ N., longitude 20° 9′ W.; wind westerly No. 5; ships head southerly; extreme rolling 20° each way, 11° commonly; card A swinging 3 to 1 points,

March 20th to 22nd. Whilst at anchor at Porta Praya, Island of

* Compasses on this plan are made by Mr. W. Heath, Mathematical Instrument Maker, Fore Street, Devonport, who has power to do so by my authority.—W. W.

St. Jago, riding with a fresh N.E. trade wind, the ship's head would sheer with gusts on the bow; and it was only by watching very closely that any bearings could be taken. The cards and pivots were changed occasionally.

April 4th. In the S.E. trade ; heading up S.b.W.; ship steady; card D steady.

April 19th. Latitude 31° 14' S., longitude 8° 22' N. ; light S.Wly. winds ; heavy swell from the S.W.; steering S.E.; ship rolling 21° each way, commonly 10° to 15°; card D, extreme swing 3½ points, commonly 2 to 1½ points. The glass of the lamp fell out ; the putty being cracked by the heat, dropped off; fixed the glass with four studs. Not sufficient outlet for the smoke. The outlets also requiring more protection from the wind, as when the lamp is on the weather side, the light is flared and goes out. Did away with one burner, and reduced the remaining wick to ten threads ; very faint light indeed.

April 21st. Latitude 31° 45' S., longitude 1° 51' W.; wind southerly No. 6 ; close hauled ; heading up, E.S.E. ; card C swung one point in the lurches.

April 24th. Latitude 33° S., longitude 5° E.; wind W.N.W. No. 7 to 8 ; B.C.V. and Q.C.P. steering S.E.b.S. running with a heavy sea, speed 11 to 13 knots ; card C swung rapidly *from east to south, card D from E.b.S. to S.b.E., card A from east to south.*

In firing great guns the cards were jerked by the concussion, and veered round quickly; it was necessary to screw the card up.

From the Cape of Good Hope to the Mauritius, the cards were much steadier, and continued so until our arrival at Ascension, when they were observed to swing, and as we advanced to the northward they became unsteady. The cards and pivots, as previously remarked, were changed occasionally, and the gimbals were examined as to the free working of the bowl.

August 13th. Latitude 7° N., longitude 19° 24' W. ; wind S.S.W, No. 5 ; ships' head north ; S.Wly. swell; card A swung two points, least swing ½ point; tried card D, having fitted it with two thin copper bars across the needles; extreme swing one point, least ½ point; detached the copper bars, result the same.

September 1st. Latitude 35° N., longitude 27° W.; fitted card A with four copper bars, of the same size as the needles, and fixed them across with slits : sufficient opportunity has not occurred for trying their effect, the weather being so remarkably fine.

While at anchor at St. Michael's, the rollers set in, and the ship rolled 12°; the card A. then swung one point. In firing salutes, the copper bars certainly had the effect of steadying the card ; the jerk was trifling, and swing not over ¾ point. It would probably be of service to give them a trial in bad weather. The bars will be returned in the card boxes. Card A as above fitted, remained in use up to our arrival in England and was steady.

In conclusion : the cards appeared steadier in the southern than in the northern hemisphere; probably from the circumstance of the ships' head being principally to the eastward or westward , at which points the cards were observed to be more steady.

That the cards in fine weather, sailing in smooth water, are invaluable, when those usually steered by are very sluggish.

That in running before the wind in a gale, as noticed on the 24th April, (25th also, see log,) the ship rolling deep and plunging, the cards were found to be *useless ;* and had there been none others on board, the ship *must have been brought to the wind* and steadied, to ascertain any bearing.

The needles are subject to rust, and have been cleaned three times ; it is however of a light powder formation, and is easily removed.

Card A has 8 needles.

   ,,   D ,,   4   ,,   3¾ inches apart.

   ,,   C ,,   4   ,,   1¾ ,,   ,,

The lamps for steering purposes would require remodelling altogether.

HENRY DAVY,
Master.

H.M.S. Thunderer, off Scilly,
September, 21st, 1843.

The compass committee, in directing these powerful compasses to be made, overlooked the circumstance, that by increasing the magnetism in an eight-fold ratio, they increased the dipping tendency of the machine ; and that by lowering the centre of gravity, the magnetism of the eight needles would trip their common centre of gravity on one side of their common pivot of support ; in a word, that the accumulated magnetism of these compasses, united under a single card, would make a more sensitive and a better compass, for fine weather and smooth water, than ever was produced before ; but for storms and tempests at sea, when good compasses are most required, the improved compasses would prove defective or inefficient.

§ 49, Having been directed to fit a compensation regulator on my plan in 1843, to the under side of one of the standard compass-cards, having eight needles under it, I deprived these needles of their magnetism, and adjusted the *matter of the card,* so that the plane of the card was made to be horizontal, and the vertical spindle to pass through the point of percussion, centre of gravity, and point of suspension, in a line perpendicular to the plane of the card ; in a word, the material particles composing the card and its appurtenances were in perfect equilibrium around the pivot upon which it was intended to traverse. The compensator being now firmly fixed, the needles were re-magnetised and the following comparative experiments were made with an un-rectified, but otherwise a similar card :—

The compass and its card, mounted in the usual way, on being carried about and rather roughly handled, by being made to imitate the motion of a vessel, diverged from its meridian 80 or 90 degrees (by reason of the unequal weight of the opposite parts of the card) and came to rest on its meridian after a lapse of 200 seconds of time : whereas the rectified card, when submitted to similar treatment and similar deflection, diverged from its meridian about 6 or 8 degrees, and settled on its meridian after 20 seconds of time. Now, in the *first case,* the magnetism of the needles interfered with their gravity, but in the second case, these disturbing forces were separated, and the directive force alone acted upon the needles, and compelled them to adhere to their meridianal direction.

Two of the ordinary binnacle steering compasses were procured from Woolwich Dockyard, by order of the Board of Admiralty. The needles of these compasses, although found to be of steel, were so badly tempered that much of their magnetism had departed from them. They were properly hardened, and after being re-magnetised were found to have acquired additional power. One of these needles was refixed to its proper card, with its bars and *balance weights;* and the other needle was fixed to its card, from which the brass bar and balance weights were removed, and a compensation regulator *added.* The card, &c., having been adjusted in perfect equilibrium upon its pivot, and a double suspension fitted, in the way I have described, the magnetic energy of these needles was such as to support, by magnetic attraction, their respective weights; that is to say, the north pole of one needle being applied to the south pole of the other, it readily raised it, and held it by suspension.

The cards were now mounted in their respective bowls, and the following comparative results were obtained, similar pivots being used,—

Common card, single needle, deflected 135° made 51 oscillations in 458''
Rectified card,   „   „   „ 15   „   134''
Ditto   „ by the double suspension   „   8   „   60''

The following trials were made at the Admiralty, on the 24th of January, when the needles of the respective compasses were deflected 90 degrees from their meridians, and the number of vibrations counted before they came to rest: viz.—

A standard compass of the first order, with 8 needles, made 21 oscillations.
A rectified card of the same kind, in the same box, 8 needles 8   „
A rectified steering card, 1 needle single suspension 12   „
The same description of card, double suspension............ 6   „
The standard compass came to its meridian within half degree, and the steering compass one-eighth point, in a quiet room in the Admiralty.

It appears, then, from these experiments, that when we prevent the magnetism of the compass-needle from deranging the centre of gravity of the card, in the way we have shewn, we cut off all tendency in the needle to depart from its meridian by mechanical means; we prevent any mechanical tendency in the card to a rotatory or oscillatory motion; and we compel the needle to obey the only force we wish to act upon it; namely, the horizontal magnetic and directive force. which the earth's magnetism exerts upon the needle, to keep it in the direction of the magnetic meridian, and thereby enable the mariner to see by his compass, whenever the direction of the ship's keel diverges from the course he is ordered to steer.

§ 50. Opinions still prevail, that in order to produce a compass that shall be steady under extraordinary circumstances at sea, additional friction should be added; for we find in Captain Johnson's book, at page 80,* these remarks: "It would seem that the question involves itself into one of additional friction." Now, if such had been the case, it is fair to infer, that a compass committee, patronized by the Admiralty, and making compass experiments at the

* Practical Illustrations of the necessity for ascertaining the Deviations of the Compass.

public expense, for a period of ten years, might have settled this point, and produced a compass for posterity to copy; or else have given up the attempt as hopeless. But from what we can gather from official reports from H.M. ships, the powerful standard compasses supplied from 1843 to 1853, furnished as they have been with two or three kinds of cards for each instrument, are still as unsteady in storms as ever. The weight of some of these cards has been increased so much, that I have met with several having had their agates or rubies broken or drilled through, by motion combined with pressure upon their vertical pivots.*

Binnacle compasses, delicately mounted and furnished with single-edge bar-needles, of considerable magnetic power, are all equally faulty in a rough sea. We have elsewhere remarked, that the weight required to counterpoise the needle's tendency to dip, becomes a measure of the needle's magnetism; and the greater the magnetism of an unrectified needle, the greater its tendency to oscillation in stormy weather. Now, the many compasses that have been fitted on my plan, and sent to sea in H.M. ships, have been reported favourably on, as steady in stormy weather; and I have never heard any of them mentioned as being sluggish in smooth water. These instruments have a double suspension and a double capacity for facilitating horizontal motion; and yet their steadiness in stormy weather is ascribed, by the committee, to additional friction. " It would seem that the question involves itself into one of additional friction." The cards of my instruments are light, and the needles, from form and mode of tempering, are very powerful indeed, as compared even with other needles of the same weight and length.†

§ 51. By an order from the Admiralty, dated 16th October, 1845, I superintended the forging and magnetising of two dozen of compass-needles, made on my plan, from the ordinary cast steel, served in by contract to the Dockyard at Devonport. These needles were heated in *boiling lead*, and cooled or tempered in boiling water; consequently they were made equally hard throughout. These needles were to be tried, for comparison of magnetic powers, with other needles of the same length and weight made elsewhere, and sent to the Dockyard to be reported on by the proper authorities to the Admiralty.

---

* The author examined compasses and swung H.M. ships, at Plymouth, as a part of his duty.

† In H.M. iron steam-ship Trident, when in command of Lieutenant Risk, they experienced great difficulty in steering the vessel, by reason of compass oscillations, when the vessel rolled much on a northerly or southerly course. The commander applied for, and was supplied with a binnacle compass on my plan, with a single but powerful needle, fitted to a light card. On going between Plymouth and Liverpool, Lieutenant Risk informed me, that my compass was not steady on a north or south course, with a sea causing the vessel to " knock about." I explained to him, that this arose from the great power and sensitive mounting of the compass, obeying the successive impulses of the iron ship's magnetism, as it passed from side to side as the vessel rolled in a sea-way. In order to remedy this evil, I took this compass-card into the commander's cabin, and, by means of deflection, shewed him the great magnetic directive power of the needle, as compared with other compass-cards supplied to the Trident. I then, by means of magnets, took away half the magnetism from my needle, making a notation thereof. On the return of the Trident to Plymouth, 4th December, 1846, Lieutenant Risk informed me, that my compass " was still sensitive enough in smooth water, and much steadier in bad weather, when steering northerly or southerly courses." I prefer this kind of evidence to hypothetical opinions.

The experiments were accordingly made in my presence, and the following table exhibits the result obtained :—

Compass Needles on Mr. Walker's plan, made from common Cast Steel, served into Devonport Dockyard.								Compass needles made, fitted, & magnetised by Mr. Cox, Optician, Devonport.			
Number.	Length.	Weight.	Deflection at 2 lengths.	Number.	Length.	Weight.	Deflection at 2 lengths.	Number.	Length.	Weight.	Deflection at 2 lengths.
	in.	gr.	°		in.	gr.	°		in.	gr.	°
1	6½	625	17	12	6½	625	18	1	6½	625	14
2	„	625	15	13	„	625	17	2	6⅜	680	14
3	„	620	18	14	„	630	18	3	6¼	780	14
4	„	620	18	15	„	630	15	4	6¼	725	13
5	„	625	16	16	„	630	17	5	6¼	755	10
6	„	620	17	17	„	625	17	6	6¼	750	14
7	„	625	18	18	„	630	15	7	6¼	780	17
8	„	630	17	19	„	625	19	8	6¼	785	16
9	„	630	19	20	„	625	18	9	6¼	685	13
10	„	620	18	21	„	635	18	10	6¼	735	16
11	„	625	18	22	„	625	16	11	6¼	735	16

Number 1 of Mr. Cox's needles was a *pattern needle*, sent into the Dockyard as a guide for the workmen, by which pattern, two dozen were to be made of the same length and weight. Mr. Cox sent in only *one dozen*, instead of *two dozen*, for trial and transmission to the Admiralty. The reader, on looking over the table, will see in the columns of *deflection*, the magnetic and directive power of each needle in proportion to its weight and length, &c. The whole of Mr. Cox's needles were made *shorter* and *heavier* than his *pattern needle;* and yet their magnetism by deflection, appears inferior to lighter needles, tested at a greater distance from the deflected compass.

A needle of the same weight and power as No. 20, when fixed to its card and cap, &c., weighed 1080 grains, and was sent on board the *Recruit* as a binnacle compass, on my plan, for trial, with about 15 other kinds of compasses. I weighed the *Recruit's* standard compass-cards, and compared their deflecting power with mine, and the following is the result; viz.—

1st. Mr. Walker's card.......... 1 needle, 1080 grains, deflection 18° at 2 lengths.
2nd. Admiralty card in Recruit, 8 „ 1570 „ „ 17° „ „
3rd. Ditto ditto ditto ditto, 12 „ 2460 „ „ 23° „ „

These compound needles for the *Recruit's* standard compass were made of Strasburg steel, manufactured in a particular way, for being fitted as main-springs for chronometers. It will be evident, by reference to the above results, that the directive power of the compound needles, as compared with that of the single needle, No. 1, are really small in proportion to the weights they have to carry; and yet, in this vessel, these heavy cards were among the most unsteady in the brig in bad weather; whilst my powerful but light card, with its compensation for dip, and its double suspension, was the steadiest as a steering instrument, and was *sought for by the quarter-masters and helmsmen*, when other experimental compasses failed them at night. The question, then, does not involve itself into one of additional friction, but is dependent on the principle I was the first to explain and demonstrate to the Lords Commissioners of the Admiralty, a principle which compass makers and the Admiralty Committee are unwilling to admit, and ashamed to deny or publicly discuss.

§ 52. Experience has proved, that our earth gives out, or imparts magnetism to every thing upon its surface, just in the same way that magnets communicate their magnetism to bodies capable of receiving it, when placed within their sphere of action. Iron of every kind has, of all known substances, the highest capacity for receiving magnetism, and hardened steel has the greatest capacity for retaining it.

We have shewn, (§ 21,) that a cast iron shot or sphere becomes pervaded by magnetism received from the earth inductively. That the shot has a north and a south pole, with a magnetic equator, where the attraction = 0. That the ball resembles a little world, and acts magnetically upon a small delicate compass-needle, by attraction or repulsion, just as we please ; for in changing the position of the shot, we alter the position of its magnetic poles, as well as of its magnetic equator.

If the directions given at § 21, *fig.* 4, be followed, and a line be drawn, or a thread tied round the ball, parallel to A B, or at right angles to the dip C D, then this line A B will represent the magnetic equator, and will divide the ball into two magnetic hemispheres; so that every part of the ball above the line A B will attract the north end of a small compass-needle, and under the line the north point will be everywhere repelled. I know of no experiment that so clearly demonstrates the nature of induced terrestrial magnetism as this one. The reader will bear in mind, that every article of iron (cast or wrought), whatever its form or magnitude may be, will be magnetic by induction from the earth, and have two poles of opposite kinds, just like our shot ; and whose axis will be parallel to the dip at the place of observation or experiment.

It is on this account that an iron ship exhibits such powerful effects upon her compasses. A floating iron ship may be regarded as an inductive floating magnet, liable to have its poles changed by every alteration of course, or of every change in the inclination of the vessel, or transfer to another magnetic latitude. The reader, in looking back to § 21, and *figs.* 1, 2, & 3, will observe that the bolts or bars, placed in different positions, act differently on each other, or on a small magnetic needle ; but these bars are made magnetic by the earth, in the same way as the iron sphere is magnetised by it. The positions of the induced poles of any regular mass of iron will always be referable to the magnetic dip at the place of observation ; for example, on the earth's magnetic equator, the polar axis of the iron in a ship, instead of approaching the vertical, as in high latitudes, would actually be in a horizontal direction, and parallel to *the magnetic* meridian.

There are upwards of 1000 tons of iron employed in the construction and equipment of a first-rate wooden sailing man-of-war,
and many merchant vessels carry cargoes of that metal, stowed in all
kinds of ways.   Need we be surprised, then, that the steering
compasses are deranged by the magnetic action of ships and their
contents?   But how much more will compass deviations be augmented in our great iron steam-ships, and in our armour plated ships
made impenetrable to shot or shells?

§ 53. When a ship is upon the magnetic equator, her compass-
needle has no tendency to dip; and when near the magnetic pole,
the needle, however free to move, would have no tendency to
horizontal direction: it would tend to point perpendicularly downwards, and be useless for purposes of navigation.   We seldom hear
complaints of compasses near the magnetic equator; for there the
magnetic line (dip), coincides with the magnetic meridian, and all the
iron in a ship will have its poles in a direction parallel to the compass-needle, or nearly so.   But when ships advance into high latitudes, the dip increases; the directive energy of the compass
becomes less and less; whilst the induced magnetism of the ship's
iron increases, and its disturbance becomes greater, and the deviation of the compass or its local attraction is augmented.

Powerful compasses without rectification require additional weights
to preserve horizontality in the cards; and this apparent balancing
actually increases the difference between the weights of the north
and south semi-circles of the compass-cards; the balance weights
making the difference between them: the ship as she rolls or rocks
from side to side, carries her compasses through the air, alternately
in opposite directions, and the north and south sides of the needle
and card being actually made of unequal weights, a mechanical
oscillation of the compass-card, in a horizontal plane, arises, which
seamen might with propriety compare to the swinging of a topsail,
or topgallant yard, when the braces are let go, and one yard-arm is
made heavier than the other by a man upon it.   Another great
inconvenience arises to produce errors and oscillation in a compass,
from the induced magnetic poles of the iron in a ship passing from
side to side, by the inclination or rolling of the vessel.   This disturbance is greatest when the course is north or south, and least
when steering east or west.   This agency has already been noticed,
and demonstrated in § 25, by suitable diagrams.

The mariner, then, has two disturbing forces to contend with in
the use of his steering or azimuth compass: 1st, The mechanical
error arising from the centre of gravity of the needle being drawn
by the earth's magnetism on one side of the pivot of support, (§ 48,
fig. 2).   2nd, The change which is constantly taking place in the
position of the induced poles of every article of iron in the ship, as
she changes her latitude, course, or seat in the water.

With regard to the mechanical error, it has been shewn, (§ 46,)
that this has been got rid of.   But some may argue, that the very
small weights that are really required for a compass of the ordinary
construction, to counterbalance the tendency of the needle to dip,
are altogether unimportant and insignificant.   It may be said in

F

reply, if these sliding weights are unimportant, why apply them? The fact is, they are important; they must be of sufficient magnitude to preserve the card in a horizontal position; and they are actually found to cause considerable oscillation in high latitudes: whereas near the equator no inconvenience is experienced in any compass. A powerful compass-needle, magnetised to saturation, will require six or eight grains near its ends, to counteract the tendency to dip; and if any sensitive needle be moved alternately in an east and west direction, by hand or otherwise, the needle *will* oscillate from its meridian, and we can predict the direction taken by either of its poles: for in any latitude whatever, the heaviest end of a compass-needle will be opposite the dip; and this end of the needle, by the well known laws of mechanics, will first resist motion: but having once acquired motion, the heaviest end will be disposed to go on at the end of the motion.

§ 54. About twenty years ago, I made a sectional model of a frigate, with iron guns run out in their respective ports, on each side of a small compass, placed on the quarter-deck. The model itself was entirely free from iron, and was intended to shew the way in which a compass was acted on by the magnetism of the guns alone, either as producing a permanent error in smooth water, or an oscillation in a sea-way. This experiment was made at the Trinity House and at the Admiralty: it was convincing and conclusive. At the request of Admiral Sir George Cockburn, I shewed how this oscillation was prevented, by placing a piece of soft iron in such a position as to have its poles changed with the rolling motion, and to operate in such a manner as to counteract the effect of the guns. In rolling the model, with its guns in an east and west direction, the compass remained perfectly steady. An Admiralty order was immediately sent to the Port-Admiral at Plymouth, that I should prepare a pendulous apparatus* for counteracting the magnetic disturbance of the guns, &c., of the *St. Vincent*, then lying as flag-ship at Portsmouth, but without tanks and many of her guns. The harbour of Portsmouth is not deep enough for such ships to lie afloat in with everything on board; and I suggested to their lordships, that an experiment made under the conditions referred to would not be of much scientific utility; that errors and other conditions should be ascertained, before a remedy could be provided; and I named the *Caledonia*, of 120 guns, as a proper vessel to experiment upon, she having her full equipment on board. An order was accordingly issued to try the amount of the *Caledonia's* magnetism when inclined each way; but she was ordered to sea, and this experiment was tried in the *St. Vincent*, on her arrival at Plymouth, in September, 1844.

This interesting experiment, although originating with me, was not made in the way I should have made it. It was made under instructions issued to the superintendent of the compass department, who had orders to communicate with me on the subject. I of course attended to the experiments, recommended some to be made, and noted every thing: and although all was not done that ought to have

* The pendulous apparatus was simply a rod of iron hung vertically by a bit of string, so that when the model heeled to starboard, the upper end of the rod went to port and prevented all oscillation or compass deviation.

been done for the advancement of science, we obtained sufficient data to shew, that the principles I had previously propounded were borne out by the experiments made in the *St. Vincent;* namely, that the changeable polarity in the ship's iron would be apparent as the ship was either turned round, or inclined on each side from an upright position.

We will now describe the nature and extent of these experiments, which, being made in a ship of three decks, with every thing on board, are not devoid of interest. The *St. Vincent* mounted 120 guns; had a complement of 1000 men on board, with three months' provisions, and 410 tons of water stowed in her hold; she had 807 malleable iron tanks, and 263 tons of cast iron ballast. The ship was under sailing orders, and all her stores were complete.

The steering compasses of all ships are necessarily placed near the helmsman, at the stern of the vessels. The compass then is above and abaft the great masses of iron in a ship; hence it results, that the north point of the compass-needle is, in our hemisphere, drawn forward by the magnetic attraction of the ship's iron; whilst the south end of the needle is driven aft by repulsion, ($ 21). Our object was, first to discover how much, or through how many degrees the *St. Vincent's* compasses deviated, as the ship was turned round in a perfectly upright position, with every thing in its proper place, as if the ship had been at sea. This operation had nothing new in it, being similar to the usual process pursued for determining the compass deviations generally. The correct magnetic bearing of a distant object is first ascertained. It should be sufficiently distant, so as not to have its bearings sensibly altered by a ship turning round in a circle, of which her own length is the radius. In the case of the *St. Vincent*, we selected a high hill or tor on Dartmoor, as our object of standard bearing, distant from us about twelve miles.

$ 55. The ship's head being placed upon the magnetic meridian, or in a north and south direction by the standard compass, she was gradually turned round, carefully stopped, and firmly held in the direction of every one of the two-and-thirty points of the compass, until observation and comparison was made of the *correct*, and the compass bearings of the aforesaid tor on Dartmoor. The result of this experiment proved, that the greatest amount of local attraction of the *St. Vincent*, upon her steering compass in the binnacle, did not exceed *half a point*, when perfectly upright.

This kind of deviation, in wooden ships, is almost always greatest when the ship's head is nearly east or west; and it diminishes or entirely disappears at north or south. A deviation of half a point, although apparently small, is yet of too great amount to be neglected in navigation. A large ship like the *St. Vincent* might easily run 120 miles from sunset to sunrise; and a deviation from her course, of half a point, would give an error of twelve miles in her dead reckoning.

It may excite surprise that a three-decker, containing at least 1000 tons of iron, should have her compasses so slightly affected; when we meet with three times the amount of deviation in small vessels, with very little iron about them. The fact is, that in large vessels, their huge dimensions enable the mariner to place his steering compass at a *greater distance* from the iron than he can possibly do in a small craft.

In the experiments under consideration, the *St. Vincent's* standard compass was 18¾ft. from the nearest carronade, and 53 feet from the nearest tank in her hold. Magnetic attraction varies with the distance. The force it exerts is found to be inversely as the square of the distance from a gun, &c.; so that should a gun, at 18 feet from a compass, cause a deviation of only one degree, it would, at half that distance, or 9 feet, produce a deflection of 4 degrees ; and at 4½ feet from the binnacle (not an unusual distance in small vessels), 16 degrees. We need not, therefore, be surprised that a large ship enables its navigator to remove his compass so far from the iron usually stowed in her, that the amount of compass deviations is less than that in smaller vessels.

The next object we had in view was, to find if an inclination of the vessel to either side would exhibit any difference in the magnetic compass bearing of the distant object on Dartmoor, as compared with the bearings taken on each point, when the ship was upright; that is, if the magnetism of the ship would be transferred from side to side by the inclination. The guns and other things were secured in their places, as at sea ; one hundred tons of water, in wooden casks, were procured from the shore, and placed upon the larboard side on the main-deck. This water, with the men's hammocks, inclined the ship 8½ degrees to port. The ship's head was placed at north, and we found that the muzzles of all the guns on the starboard side attracted the north point of the compass, and the breeches of all the guns on the larboard side attracted the same point : thus shewing clearly, that the north polar magnetism of the guns had moved along their length to the highest part of each (§ 25). On going down into the hold among the iron tanks, which were stowed in two tiers and in close contact, we found that all the tanks on the starboard side of the ship attracted the north point of the compass, and those on the lowest or opposite side of the ship repelled the north, and attracted the south end of a compass-needle. When a delicate compass was placed amidships, on the square of the main hatch-way on the orlop deck, it exhibited an easterly deviation of 1½ point ; the north point being drawn to starboard and the south point to port, by the magnetic agency of the tanks, and in an opposite way to that produced by the guns. The ship was then swung to each point successively round the compass, and observations were made upon compasses placed as follows : one on front of poop, one in each binnacle, and one in the bread room. The compass in the bread room had previously been placed in the orlop, but it gave early indication of being powerfully influenced by the spindle of the capstan or tanks ; and it was removed (not at my request), to the bread room, at the greatest possible distance from the disturbing agency.

§ 56. The result of this experiment of swinging the ship, when inclined 8½ degrees to port, did not exhibit any *great amount of change* in the bearings by the poop compass ; but it was evident, that the amount of compass deflection, arising from the ship's inclination, was due to the *difference* between the varying polarity of the ship's wrought iron tanks, and her cast iron guns, &c. For all the guns attracted the compass towards the *lee side*, and the two tiers of tanks, being in contact with the iron ballast and each other, acting as one mass, attracted the same point towards the weather side; the magnetism of the tanks having the mastery over the guns (§ 44). At section 33, we theoretically discussed the effect that a change of wind might have upon a ship's *compass deviations*, although her course remained constant any where between the cardinal points of the compass; and that by the force of the wind on her sails, the ship, by a change of inclination, might even cancel her upright compass error on one tack, and double it on the other; the actual direction of the ship's head remaining unchanged. Now, the experiments made in the *St. Vincent* proved the correctness of previous conclusions.

For example, when the ship's head was north-east by the standard compass, on its pillar on the poop, the ship's head was N. 41° E. by the binnacle compass (when on an even keel, or upright), and at N. 46° E. by the same compass when the ship was inclined 8½ degrees to port. When the ship's head was south-west, or in an opposite direction, the binnacle compass in the upright position gave S. 49° 30′ W., and in the inclined position only S. 47° W.; so that in the first instance the *incline deviation* was greater than the upright deflection; and in the second case it was less.

The quantities we have quoted appear small, and practically would be of small consideration, so far as a day's dead reckoning is concerned; but the results obtained are nevertheless important, as developing a principle which neither Act of Parliament nor order in council can invalidate. I regret that the superintendent of the compass department did not deem it necessary to have the *St. Vincent* inclined the other way, and swung round with an inclination to starboard.* He said his orders were to *swing the ship round*; and having done it in one direction, with one inclination, he deemed his orders fulfilled. Scientific considerations required that this huge three-decker should have been turned round with an inclination both ways. It was not done; but the tabulated results obtained, enable me to give a cursory view of the induced polar magnetism of the ship having undergone a change by her inclination; and if no other experiments had been made, these tabulated results would have been inserted here. There are, however, more convincing and satisfactory experimental results obtained, which will appear in the sequel. I will now sum up such evidence as may be worth recording, relative to the magnetic trials made on board the *St. Vincent*, in September, 1844:—

1st. That, by the ship's inclination, the induced polarity of the tanks and iron ballast passed from side to side; the highest side attracting the north end of a compass-needle, and the lee or lowest side *repelling* the north, and attracting the south point.

2nd. That the 307 iron tanks (capable of holding 410 tons of water), resting upon 262 tons of cast iron ballast, did exert as much

* He was very anxious to get rid of the experiment altogether.

magnetic influence upon the ship's compasses as 3268 tons of iron in the hold; for an iron tank acts magnetically as a solid.*

3rd. That, although the nearest gun was only 19 feet from the standard compass, and the nearest tank in the hold 53 feet from it, the magnetism of the tanks and ballast, deflecting the compass in one direction, was greater than the magnetism of the 120 guns, bolts, knees, and other things acting upon the ship's compass in an opposite direction.

4th. That the actual compass deviation in the *St. Vincent*, as demonstrated by the experiments in her inclined position, was due to the difference between the magnetic action of the guns or bolts and the tanks and ballast.

Since the distant tanks, resting on the cast iron ballast, exerted a greater influence on the ship's compass, in her inclined position, than the action of all the guns, &c., by drawing the north point to windward, and thereby obtaining the mastery over the guns' magnetism, as demonstrated in § 25; I think it will be admitted, that an augmentation of guns, or a diminution in the tanks, would bring these induced magnetic antagonastic forces, acting on a ship's compass, nearer to an equality; and thereby render a compass more steady when a ship is rolling.

We may also fairly assume, that if a line-of-battle ship landed her guns, the *incline deviations* of her compass would be augmented, and her compass would be rendered more unsteady in stormy weather. These deductions appear to me to be perfectly legitimate; but let us look for corroborative evidence.

§ 57. H.M. ship *Thunderer* was ordered to land her lower-deck guns, in 1843, and convey a regiment to the Mauritius. She was supplied with a powerful standard compass, fixed on a pillar on the front of her poop; and this compass had cards of different kinds; and some additional contrivances to fetter the oscillations, if necessary. Many of her cabin and upper-deck guns were struck down into the orlop, so that the *Thunderer's* remaining guns would act but slightly, as she rolled from side to side. This ship had also two tiers of wrought iron tanks, stowed above her iron ballast; and being a ship of two decks, (instead of three,) her standard compass was fixed a deck lower than in the *St. Vincent;* consequently the two-decker's poop compass was nearer to her tanks than the poop compass of the three-decker; and therefore, the magnetism, or local attraction of the *Thunderer's* tank, as a disturbing agency on her compass, would be very considerable, and greater than that found by experiment in the *St. Vincent.* In steering a northerly or southerly course, there would be *two* disturbing forces to produce oscillation: 1st, The mechanical errors of the compass already mentioned (§ 43), whereby the heaviest or loaded end of the needle would, at every roll of the ship, swing towards the lee side: 2nd, The opposite end of the needle, by the ship's induced magnetism, would be attracted by the magnetism of the highest or weather side of the ship, or weather side of her tanks in the hold. In our northern hemisphere, we have *seen*, that the north point of the compass-needle would be attracted

* Barlow on Magnetic Attraction.

(as in the *St. Vincent*), towards the highest or weather side, and the south point of its needle, with its load, would be dragged by its magnetism, and mechanically driven by its load, towards the lee side; and all this by a constant succession of impulses, as the ship rolled from side to side; these two forces combining to produce an unusual oscillation. When the ship approached the magnetic equator, these disturbing forces would diminish or vanish in a short time; but after passing the line of no dip, and advancing into south magnetic latitude, the polarity of the ship's iron would undergo a change (§ 31), the dip being south, and the disturbing forces we have described would be *reversed*, but equally powerful in opposite but corresponding dips.*

When the *Thunderer* returned from the Isle of France to Plymouth in 1843, I was kindly furnished with a copy of the officer's report and observation upon the working of the standard compass, which was constructed under the auspices of the compass committee. By that report, which I insert, (§ 48), it appears, that in latitude 44° N. whilst the ship was rolling 20° on each side of an upright position, and steering a southerly course, the standard compass oscillated 68°. That in latitude 33° south, whilst the ship was scudding in a S.E.b. S. direction, in a heavy sea, and going about 12 knots an hour, the standard compass oscillated as much as 90°, and became *useless* as a steering instrument: whilst the old common and weak single-needled dockyard compass, from its absolute deficiency of magnetism, was tolerably steady. The report further states, that the *committee compass* was more steady in the southern than in the northern hemisphere, "probably from the circumstance of the course being generally to the eastward or westward."

Having written to the Admiral on the subject of these compass oscillations, I was *ordered* to present myself, and explain the cause thereof at the Admiralty, I accordingly went to London, (§ 49); but previously-received opinions in scientific quarters were looked upon with favour, whilst my explanations were regarded as doubtful or unimportant. It was then that the additional friction principle began to be entertained by the compass committee.

In 1847, the *Asia*, 84 guns., a sister ship to the *Thunderer*, was fitted to bear the flag of Rear-Admiral Hornby. She had steering compasses, as well as a standard compass, on the committee's plan. Encountering bad weather in the Chops of the Channel, her compasses were found to be useless, and the Admiral had to write on the subject.† We can now explain how these powerful compasses become useless in stormy weather, when in reality a good instrument is most wanted.

In 1844, the *Caledonia*, 120 guns, commanded by Captain Milne, now Admiral Commanding on West Indies and North America, was on a cruise at sea. Captain Milne watched the action of the *Caledonia's* compass oscillations, and he informed me, that the north point of her compass-card went towards the weather side, at the end of each roll: whence we may infer, that the *Caledonia's* compass

* See Tables of Erebus and Terror, pp. 31 & 32.
† She put back to Plymouth.

oscillated from causes similar to those acting in the *St. Vincent*, the *Thunderer*, and the *Asia;* and that these vibrations can only be cured by the mechanical contrivances I devised, and by magnetic compensations made by antagonistic magnetic forces, introduced by means of soft iron correction.

§ 58. In the year 1846, an iron sloop of war, named *Recruit*, was commissioned in the Thames by Commander (now Capt.), Adolphus Slade. Her compasses were adjusted at Greenhithe, by Captain Johnson, who, as usual, supplied a table of deviations for each of the two-and-thirty points of her standard compass. When the vessel got to sea, the officers found that these deviations were inapplicable to correct the courses under sail. On arriving at Spithead, Commander Slade applied to have his compass re-adjusted, and the Admiralty directed that the assistant to the superintendent of the compass department should proceed to Plymouth in the *Recruit*, there to swing and re-adjust for compass deviations.

On the passage down Channel, with foul winds and hazy weather, difficulties were experienced in finding the vessel's position; and these difficulties being mentioned to me by Commanders Slade and Strange, I wrote, through the Admiral-superintendent, to the Admiralty, adverting to the rather unsatisfactory experiments made in the *St. Vincent;* and requesting that, since difficulties had arisen in navigating the *Recruit*, and as she had to be re-swung, I might be allowed to determine the *nature* and *amount* of her compass deviations when *inclined*, as well as when *upright*, and to have these operations performed in *my own way*, without unnecessary interference.

I obtained the sought for authority,* and the vessel was moved into Barnpool, after being re-stowed, when she was swung by me in three positions.

It was predicted in high quarters, that these experiments would be disposed of like those of the *St. Vincent*, and that the subject deserved no serious consideration. The result of the experiments, however, commanded attention. Commander Strange was sent down to repeat the experiments, and their Lordships ordered subsequent and similar experiments to be made in several iron ships, by Captain Johnson, who has recorded some of them in his valuable work, published in 1847. The attraction and repulsion of an iron ship's sides are seldom alike; that is to say, the magnetical and mechanical axes of the ship, when her head is north or south, do not coincide, but *cross* the direction of the ship's keel. In the *Recruit*, this difference was about 5°; in the *Birkenhead*, it *was* 7°; in the *Styx* and some other vessels, it amounts to 6 or 7°, and sometimes to 8°.

The following table of the *Recruit's* deviation under different conditions, and at the table page 44, clearly prove that the compass points of no deviation have in every case changed considerably.

These changes in the 'resultant points' arise from the variable inequality of the magnetism of the sides of the ships. The points of no errors in the *Birkenhead*, when in England, were near N.N.W.

---

* I had now full power to try the experiment fairly and the result proved that errors arising from a ship's inclination can never be neglected without danger. W.W.

and S.S.E.; but at the Cape, her compass would have errors on these points. Before the ship struck on the rocks off Point Danger, she had been steering a S.S.E. compass course, as was stated by the surviving helmsman; and there is ground for our belief, that the officers who perished had placed too much confidence in the 8th practical rule for ascertaining compass deviations; which states, that the points of no deviation, when once ascertained, may be regarded as constant.

The following Table exhibits the Recruit's Compass Deviations, when upright, and when inclined 8° both ways.

Correct Magnetic Courses.	Deviations of Compass.			Deviations with Iron apparatus applied to partially correct Compass, when ship upright.
	When vessel upright.	When vessel 8° to Port.	When vessel 8° to Starbrd.	
	° '	° '	° '	° '
N.	4.15 East.	2. 0 East.	9. 0 East.	1. 0 West.
N. by E.	11.15 „	5.15 „	15. 0 „	0. 0 „
N.N.E.	13. 0 „	8.30 „	19.30 „	2.30 East.
N.E. by N.	15.30 „	10.15 „	20.15 „	3.45 „
N.E.	17. 0 „	12. 0 „	20.30 „	3. 0 „
N.E. by E.	13. 0 „	14.30 „	21.45 „	3.15 „
E.N.E.	18.30 „	16.15 „	22.30 „	1.45 „
E. by N.	18.45 „	13.45 „	22.15 „	2.45 „
E.	16.53 „	16. 0 „	16.30 „	2. 0 „
E. by S.	13.25 „	16.15 „	14. 0 „	2.15 „
E.S.E.	13. 0 „	13.10 „	9.15 „	1. 0 „
S.E. by S.	6.45 „	12.15 „	2.45 „	1.30 „
S.E.	5. 0 „	9.30 „	3. 0 West.	2. 0 „
S.E. by S.	2.45 „	7.15 „	10.45 „	2. 0 „
S.S.E.	5. 0 West.	7.15 „	12. 0 „	2. 0 „
S. by E.	6.45 „	2.45 „	14. 0 „	2. 0 „
S.	6. 0 „	1. 0 West.	14. 0 „	2. 0 „
S. by W.	8.30 „	0.45 „	16.45 „	0. 0 „
S.S.W.	9.30 „	2. 0 „	18. 0 „	0.30 „
S.W. by S.	9. 0 „	6.15 „	19.45 „	1.30 „
S.W.	9.15 „	7. 0 „	17.30 „	2.30 „
S.W. by W.	11.15 „	7.15 „	16.15 „	4. 0 „
W.S.W.	11. 0 „	9. 0 „	14. 0 „	1. 0 „
W. by S.	10. 0 „	10.15 „	12.15 „	0. 0 „
W.	8. 0 „	10.30 „	11. 0 „	1.30 West.
W. by N.	6.45 „	10.15 „	7.15 „	3.30 „
W.N.W.	8. 0 „	8. 0 „	7.30 „	4.30 „
N.W. by W.	5.15 „	6.15 „	4.15 „	4. 0 „
N.W.	3. 0 „	7. 0 „	1.30 „	3.45 „
N.W. by N.	1.15 „	3. 0 „	0.15 East.	3.15 „
N.N.W.	0.30 East.	2.30 „	4.30 „	3.30 „
N. by W.	3. 0 „	1.15 „	6.15 „	3.30 „
Total amount of Deviations.	296. 3 „	328.10 „	401.53 „	69.45 „

The *Recruit* had *wooden beams*, and iron topsides, and was armed with twelve long 32-pounder guns. On repeatedly swinging her round, we observed, 1st, that when her head was placed in a northerly direction, all the compasses abaft were more sensitive than when her head was in a southerly direction,—probably on account of the brig's magnetism and terrestrial magnetism acting in conjunction on northerly, but in *opposition* on southerly courses. 2nd, that on changing the ship's direction from point to point, a few minutes of time were required for the new compass deviation to be *fully* developed. The binnacle compass was placed before the wheel, and close to the after side of the cabin skylight, affording but small means for fixing any iron materials for correcting the compass deviations; and the means I did use for that purpose were *without any regular authority*. We were at a distance from the Dockyard, and time pressed upon us; the services of the vessel being wanted at Madeira.

Commander Slade, on receiving such a list of deviations for his compasses, enquired officially by what compass he was to keep his reckoning, or by what means he was to correct his courses;* and was ordered to adhere to those deviations found and furnished by Captain Strange, which did not sensibly differ from the deviations found by Captain Johnson at Greenhithe.

The following official report was made by me to the Commander-in-Chief at Plymouth:—

"Bevisand, 4th February, 1847.

"SIR,—In obedience to their lordship's order of the 23rd November last, 'to cause the *Recruit* to be heeled as much as may be convenient, in order to try the effect that the inclination may have upon the deviation of the compass, to report the results of the experiments, and to submit any observations I may wish to make upon the subject;' I have the honour to inform you, that the experiments have been made, and the results are exhibited in the annexed table. The ship's head was placed upon *correct magnetic bearings*, and not upon those points that a compass in the binnacle would indicate. For this purpose, the standard compass was used as a theodolite, to measure the angle that the ship's keel made with a distant object. The observers were the master and master's assistant, selected by the Commander, who, with myself, checked the observations; and in this way many sources of error were avoided.

"It will be seen by the table, that the upright and incline deviations are very different. When the ship heels to starboard, on a N.E. course, the error of the compass is greater than when upright; and when she heels to port, the error is less. In some directions, the difference may amount to a point, by a change of inclination from side to side.

"Having ascertained these facts, I prepared an iron apparatus, viz., a sphere, cylinder, and pendulum, with a view to correct or diminish the errors of the compass, which, in this iron ship, are very remarkable and complicated. I prepared this apparatus in the Dockyard, and fixed it in the brig; but was surprised to find that the iron hull of the vessel had absorbed or weakened the magnetic energy of my apparatus; so that the influence of the iron ship had greatly reduced the energy of my iron apparatus, thereby shewing that iron acts magnetically upon iron. It was therefore necessary to *augment* my apparatus, since the form and fitting of the *Recruit* does not admit of a *choice of distance*.

---

* When the Recruit left Plymouth, after being swung, on taking leave of Captain Slade, I jocosely said to him,, "If you do not actually lose your iron-aided craft, do let us hear how you get on.' I received a letter from him, dated from St. Michael's, 17th September, 1847, which began thus:—

"My Dear Walker,—I will ease your conscience, by telling you that we have found land. You could hardly have expected us to do so after having be-devilled our compasses. Joking apart our deviations have altered on many points. We are never easy when running, or making land, and have to trust to sights; but in the absence of sun, under such circumstances, I have made up my mind to trust to the binnacle compass. As eels become reconciled to skinning, so we get accustomed to faithless needles........"

The Recruit had sixteen different kinds of experimental compasses on trial, by order of the Board. My compass was placed in the Binnacle, and compasses on my principle continue to be issued to H.M. ships.

"On fixing this apparatus and swinging the ship, it was found to be an easy matter to make the compass correct, with soft iron, upon the cardinal points; but the errors were considerable upon intermediate points. My aim was ultimately to make the errors on an average as small as possible, and thus place the *Recruit* in the condition of a wooden vessel, where the errors being small, no serious damage can arise by neglecting them, or even by applying them the wrong way, as I fear is sometimes done.

"The results obtained have shewn that I was not wrong in the principle I had propounded, although I am not satisfied with the corrections. We had completed our operations, and were unable to account for appearances in the results obtained in swinging the ship *inclined to port*, when we discovered that the brig's skylight had been built, and her binnacle placed three or four inches nearer to one side than the other; and this circumstance is sufficient to account for the north point of the needle being drawn to one side of the vessel, when her head is upon a north or south line. Assuming the skylight and binnacle to have been in the midship line, I fixed my apparatus accordingly; but the original error in the skylight and binnacle has to some extent vitiated the experiments. The errors, however, are greatly reduced in the upright positions; and I am desirous of having a trial at sea, in order to settle an important principle in connection with the correction of a ship's compass. I found experimentally, about twenty-eight years ago, that the ship's iron at the Cape of Good Hope, produced a deviation in an opposite direction to the deviation in England; so that if a vessel left England with an official table of deviations, and applied these corrections to her compass during the passage, her compass might, by such aplication, be made correct at *starting*, but might be three points in error at the Cape. I have insisted that such would be the case, either by using tables of corrections as found here, or by the application of permanent magnets to correct the compass. But I am of opinion, the soft iron correction will cease to act upon the compass, as the other *iron* in the ship loses its power of disturbing it. It is, therefore, highly desirable that this point should be settled by an experiment in a ship like the *Recruit*, (a vessel of most complicated magnetism), with the apparatus just as it is; and when she returns, the binnacle and apparatus could be fixed where it should be upon a midship, and fore and aft line.

"I am, Sir,

"Your most obedient humble Servant,

"WILLIAM WALKER.

"Admiral Sir John West, K.C.B.,
  Commander-in-Chief, Plymouth.

§ 59. We have endeavoured to shew how very distrustful ancient mariners were of all those who meddled in any way with the compass; and how conservative they were in adhering to old opinions, and rejecting new ones; even when supported by experimental proofs. Highly gifted individuals of the present time, in their endeavours to generalise facts with unsound data, have now and then contributed to throw doubts upon the principles they themselves have, in other parts, propounded. Hand-books have been published for the guidance of navigators, so far as regards the compass and its deviations: artificial magnets were at one time to correct ships' compasses all the world over: an iron disc was to perform the same desirable end: the troublesome horizontal vibrations of the compass were actually to be got rid of altogether, by the adoption of compass-cards having 4, 8, or 12 needles of the greatest possible magnetic power: deviation tables have been supplied to ships for correcting their courses; and other kinds of tables have been printed for re-

forming deviation tables, should changes take place in their amount on the points of greatest deviation; these latter tables being grounded on the assumption that the points of *greatest* and *least* deviation remained constant. Unfortunately all these plans have partially failed.

Very excellent " Practical Rules" have from time to time, been published by authority of the Admiralty, for the information of navigators, relative to the local attraction of ships and their magnetic action upon the compass; but these rules have required, and will again require revision, so soon as opinions become settled on these subjects.

In the 10th " Practical Rule for ascertaining the deviations of the compass," we find the following information given (edition 1842); " Experience has shewn, that the amount of deviation on the several points varies in distant regions; and that in the southern hemisphere *it may even become westerly* on the several points on which it was easterly, and in the northern hemisphere easterly where it was westerly." In 1847, we find the words in italics of the above sentence altered as follows: " *It frequently becomes westerly.*" And in the last edition of Captain Johnson's work of 1852, there is another new wording in the 10th practical rule, running thus: " *It generally becomes westerly.*" These alterations evidently shew, that opinions or conclusions have not yet been arrived at; yet these changes of phraseology teach us to be tolerant of the opinions of others. He would be a bold man who would now stand forth and attempt to publicly vindicate the practice of correcting, by permanent magnets, the ever-changing inductive magnetism in an iron ship: for experience has taught the mariner, that his compass, corrected by magnets at home, becomes very erroneous abroad; that although his compass error at London or Liverpool may be made = 0, yet at Sydney or New Zealand its error may be = 3 or 4 points; and that instead of the errors being corrected by magnets here, they may even be doubled elsewhere.

Some fifteen years ago, one of the Niger Expedition iron ships, named the *Soulan*, had her compasses very correctly adjusted by magnets, &c., in England, and on arriving at Sierra Leone she was re-swung and as much as 12 and 14 degrees of errors were found on some points of her compass, where no errors whatever were found in England at her departure.

A gentleman who professes to correct compasses at Liverpool, wrote me in 1847: "A master of an iron vessel, sailing from this port to Bombay, told me that the compass near the wheel, corrected by magnets, was out one or two points on getting near the line ; but on coming or going northward, it gradually came back again." The iron Royal Mail Packets, that now pass with such rapidity between the two hemispheres, experience the inconvenience of permanent magnet correction. Some commanders take the magnets away from their compasses when they get well to the southward, and approach the *line*, and replace them as they return northward, and by this means avoid the doubling of compass errors.

§ 80. If a ship were supplied with a table of compass deviations *in England*, and if the navigator were to continue the application of such a table to correct his courses in the same amount of dip in the northern hemisphere; then might he double his compass errors, instead of correcting them. If a ship had one compass corrected by magnets, and another of the same kind, in the same ship, and under the same amount and description of disturbance, had its errors correctly *recorded in a deviation table;* and if the navigator were to assume that his corrected compass remained always perfectly right, by applying his recorded deviations constantly to correct the courses steered by the *other compass,* he might find his two dead reckonings agree. The iron steamer *Adelaide,* that left Plymouth early in January, 1853, had her steering compass corrected by magnets, and another in a standard position on the top of her saloon, for which a deviation table had been supplied. These compasses will greatly differ during the voyage, and the uncorrected compass will be the most to be relied upon.

Permanent magnets for compass corrections are, however, very convenient for our steamers employed in the home trade; and so long as the ships are kept *upright* or nearly so, no great danger need be apprehended; unless the magnets lose a portion of their correcting power, or some alteration be made in the ships or their machinery.

In iron vessels, the compass deviations abaft may on some courses amount to some five or six points. These great errors are generally to be found in iron screw-vessels, that have their steering wheels and binnacles near the stern, and just before an upright iron stern-post, with its iron rudder working on it. These vertical masses of iron are powerful inductive magnets *by position,* and conduct upwards the magnetism of the lower parts of the vessel (§ 24), thereby acting powerfully on the compass; attracting the north point towards the vessel's stern, and not forward, as is the case in the majority of wooden ships. Now, magnets are convenient for counteracting these great deviations, in such iron screw-ships as may be employed in the home trade; but large iron ships of any kind, intended for distant employment should not have their steering wheel and binnacle too far aft, for the reasons just given; unless means were devised for moving counteracting magnets, so as to adjust the binnacle compass at any time, as a mere matter of convenience for the helmsman. The course in all cases should be shaped and recorded in the log from a standard compass, placed in a permanent position, where there is no error, or where errors have been corrected by iron alone, or where small errors, arising from change of geographical position, or from the ship's inclination when under sail, are indicated by a deviation compass. When two compasses are placed in one binnacle, or when two binnacles in the same ship are too near to each other, they will be within their sphere of mutual attraction, and become erroneous on all courses, except north, south, east, or west. On all other courses, the compasses will be in error, arising entirely from their mutual influence on each other; and yet the two compasses, if similar, will be found to agree. Any body may try the experiment, by placing two compasses on a table, and turning the table slowly round. The compasses will agree when they bear north, south, east, or west *of each other*, but on all other points they would give a bearing of a distant object, differing from the bearing of the same object when the two instruments bear north, south, east, or west of each other. The conclusions to be drawn are, that only one compass should be used at a time, and that one should be placed if possible exactly amidships. If, however, the ship be large, and steered by a wheel of large diameter, then the position of the helmsman will not be amidships, and two binnacles may be necessary; as is the case in large ships of war, where the binnacle compasses are kept four feet apart, in order to avoid compass errors from too close proximity.

We have endeavoured to shew how the position of an irregular mass of iron may either attract or repel the mariner's compass, and divert it from that correct magnetic bearing it would indicate if acted on by the magnetism of the earth alone. This kind of compass disturbance is called, compass deviation, to distinguish it from the variation. The local deviations of a ship, arising as they do from the magnetism of the ship and her contents, are in some measure under control, and may be cancelled by counteracting iron, or by means of permanent magnets, so long as a ship continues in seas where the magnetic dip and magnetic intensity of the earth do not sensibly vary. The variation, however, is altogether beyond the control of man. It varies irregularly from time to time, in different parts of the world. There is even an annual, a monthly, and a diurnal variation of the compass, but of too small amount to influence a ship's reckoning. The variation appears to arise from some periodical and physical changes going on in the earth itself, which still remain among the unexplained mysteries of nature.

§ 61. We have endeavoured to explain, at considerable length, how a ship's compass acquires errors from the local attraction or repulsion of iron in its vicinity; and how this iron acquires magnetism and polarity by its position in the ship, and its position with regard to the magnetic dip; that these relative positions change, as the ship turns round, inclines from side to side, or passes from one hemisphere to another, or from one magnetic latitude to another. The earth's magnetic intensity is also found to be inconstant. Hence it results, that although the stowage of a ship may remain undisturbed, and the relative position of her compass, and every article of iron in her, remain constant during a voyage, her deviations and local attractions *will constantly* vary; even from *plus* to *minus* quantities, and *vice versâ*. We need not insist upon this, since the tables we have given demonstrate the fact. The local magnetic disturbances of a ship's compass are decidedly variable quantities, and any attempt to correct a compass permanently, subject to so many changeable disturbing forces, cannot be effected by a counter-application of *constant forces*, as has been attempted with permanent magnets, and found to fail.

Neither can compasses be rectified by the application of any hypothetical or unsound principles of terrestrial magnetism. Local attractions, as found existing in almost all vessels, on certain points of their courses, in smooth water and an upright position, as well as those annoying compass oscillations which are so detrimental to good steering and correct reckonings, can only be correctly cancelled or cut off by counteracting the disturbing forces, and by a proper application of materials of the same quality or kind as those that derange the free and correct pointing of a ship's compass. The disturbing and correcting materials should be of the same kind, in order that similar inductive magnetic changes may arise in the metals that disturb, as well as in those intended to correct a ship's compass, as she may pass from one magnetic latitude to another.

If space were ample, we might instruct our readers by recording instances of ships being wrecked and lives lost through sheer pedantic ignorance, neglect, or indifference to professional detail. Our aim has been to avoid, as much as possible, unpleasant recitals, and unpalatable quotations; giving only such proofs as may be necessary for supporting the views we entertain of the magnetism of ships, and its operation on the mariner's compass.

If steering compasses in ships have occasionally been made erroneous through ignorance or neglect, they may sometimes have been corrected by accident or design, in the equipment, stowage, or general arrangement of the iron on board. Brief reference has been made (§ 1,) to observations recorded by the Author on the English coast and at the Cape of Good Hope, in a vessel containing much iron of different kinds, and to a letter he wrote to Admiral Sir Byam Martin, relative to the effect that iron lower-masts would have on the steering compasses of a frigate. In that letter we asserted, that the keel and lower half of the *Phaeton's* iron mainmast "would attract the south point of the compass; the head and upper half being at too great a distance to counterbalance the influence of the heel and parts below." When the *Phaeton* was swung at Spithead, in February, 1825, and her compass deviations tabulated and discussed by Mr. Bennett, of the school of Naval Architecture, it was found that her compass errors were very small indeed; that the greatest deviation did not exceed ½ of a point (1° 40′). Mr. Bennett, in discussing the effects produced by the iron mast, says, "*that the guns, &c., had greater effect on the needle than the mast and bowsprit; so that the local attractions in the fourth column are not to be attributed to the latter, but to the former.*" And in another part of Mr. Bennett's able paper, he expresses an opinion, that if the *Phaeton* had been supplied with an iron foremast as well as mainmast, "the local attraction might, therefore, have been reduced to zero."

At page 51 of Captain Johnson's work,[*] there is a table of the compass deviations of H.M. steam-ship *Styx*, under three circumstances; viz.,

1st, Fitted as Surveying Vessel, with an iron crutch abaft standard compass.
2nd, ditto   ditto   with   ditto   removed.
3rd, Armed and stowed as a man-of-war.

The difference in these deviations is very remarkable, and shew how much the points of greatest and least compass deviations have changed places; and, as Captain Johnson justly remarks, "affords an interesting specimen of the working of the compass system which

[*] Practical Illustrations of the necessity for ascertaining the Deviations of the Compass. By Captain Edward J. Johnson, R.N., F.R.S., Superintendent of the Compass Department of the Royal Navy. 1852.

has now been adopted." That part of the table referred to which exhibits the sum of all the compass deviations in the *Styx*, with an iron crutch fixed abaft her standard compass, and the sum of the said deviations *after* the said iron crutch was removed, is deserving of especial notice. By taking away the objectionable crutch, the deviations of the *Styx's* standard compass were *increased* 33 per cent !*

The applications I made, at my own risk, in the iron sloop of war *Recruit* (§ 58), however inconvenient the vessel was for such a purpose, and however unskilful the application was on my part, the tables of deviation shew a diminution of 75 per cent. in the upright position.

Professor Barlow put in practice a plan of his, for correcting compass deviations, by means of a circular disc of soft iron. This plan actually corrected the compasses of several sailing wooden vessels, to a very great extent. The disc was to be fixed by trial, so as to cut off the maximum errors, according to instructions given. But Mr. Barlow has admitted, at page 102 of his *Essay on Magnetic Attraction*, 1824, the probable "impossibility to get a correcting plate to give the same attraction as the vessel at every point." He had firmly persuaded himself, that effects produced upon the magnetic needle depended entirely upon the position of the centre of the mass of disturbing iron ; that the action of plain unmagnetised iron on a compass was referable to two poles, "indefinately near to each other," in the common centre of attraction of the surface of the body (p. 188); and he altogether repudiates the fact, that iron, magnetic by induction from itself, has polarity or directive energy.

At page 115, he says, "According to the present received doctrine of magnetism, every mass of soft iron becomes a temporary magnet by induction from the earth; yet I am not aware that ever any particular action has been discovered between two pieces of iron. For instance, they will not, that I am aware, attract and give direction to the lightest and most freely suspended needle of soft iron or unmagnetised steel." In another part of his valuable work, he states distinctly, that if iron were really polarized, and acted magnetically on other iron, then would his theory of compass correction be inadmissable.

§ 62. We are taught to believe, that terrestrial gravitation is greatest at the earth's surface, and at the earth's centre = 0. Our experiments on inductive terrestrial magnetism lead us to infer, that the induced magnetic poles of an iron sphere, and their attractive or repulsive forces, would be found on the *surface* of the sphere, and not at the centre, as assumed by Barlow and others. We actually find that such is the case, when we apply a test compass, either within or without a hollow iron sphere; or even to a cubical iron tank or other vessel. An iron disc recommended for counteracting the local attraction of wooden ships, if placed with its plane in a vertical position, and parallel to the plane of the magnetic meridian, its induced poles would be found in England at the circumference of the disc, and on a line parallel to the direction of

* Very interesting; the crutch deemed objectionable: when removed, the Compass errors augmented 33 per cent!

the dip. But when we change the position of the disc, and place its plane in an east and west position, at right angles to the dip, its induced magnetism and polarity *appear* to have vanished altogether; and that its magnetic poles have vanished. This is, however, a fallacy; there is just as much induced magnetism in the disc as ever, but its induced poles are now to be found, if I may be allowed to use Mr. Barlow's words, "indefinitely near to each other."

For the sake of further illustration, let us suppose our circular disc to be 18 inches in diameter, and half an inch thick; then in its vertical and north and south position, its induced poles would be 18 inches apart; but when its plane is placed at right-angles to the dip, the induced poles of the disc would be only ½ inch apart. A thin disc of iron, although it might be placed in one latitude to correct a ship's compass from *upright* local attraction, would not in the same position correct it in another latitude. An iron sphere would be a better machine to accomplish such an object, since its induced poles would always be a diameter asunder.

I applied a cast iron hollow sphere (13 inch shell), in an attempt to correct or partially cancel the upright deviations in the *Recruit*, and the reader, in looking back to § 58, will see that about three-fourths of the errors were reduced. The sphere, however, could not correct or cancel the change of polarity in the ship, arising from her inclination or rolling; because the table shews clearly, that the north point of the compass-needle was drawn to leeward; whereas in the sphere, no change would be apparent on either side. An iron sphere at the magnetic equator would have its induced poles at a diameter apart, and horizontally north and south of each other, although turning constantly round any axis whatever. I have found such to be the case experimentally; and therefore it is not strictly true, that the local attraction of a ship's iron "*vanishes* at the magnetic equator." Its effect on the compass diminishes, but does not entirely vanish. The polarity of the iron and the pointing of the compass have a tendency to the same direction; but if any master or mate of a merchant vessel, having an iron tiller, or iron *spindle* of steering wheel, be placed in a horizontal direction will place a compass either *east* or *west* from the ends of such spindle or tiller, when the ship is steering a northerly or southerly course near the equator, he will soon be convinced that horizontal iron acts far more powerfully *there* on a compass, than it would do any where else on the globe.

Many ships have iron transoms, and other heavy fastenings about the stern, that may correct or derange a compass in the binnacle. In some of our sloops of war, fitted with iron transoms, life buoy irons, &c., when the standard compasses have been fitted abaft the wheel, I have found that the small quantities of iron abaft have more than counteracted the action of the ship and her contents lying before the compass. In such vessels, the actual deviation seldom exceeds ¼ or ⅓ of a point. The *Racer, Ranger, Pantaloon*, and *Heroine* are examples of wooden vessels in which the deviations were very small, and the attractions towards their sterns.

It is evident, then, that when a ship's compass is placed in the binnacle, in a fitting and convenient position, and found to have

G

errors arising from the local attraction of the ship's iron, it would be an easy matter to cancel these errors by iron alone, and without magnets of a permanent kind. A bar or bolt of iron, or a hollow tube of iron, will, by position, act in every respect upon a compass-needle as a magnet, and may be applied to correct a ship's compass, according to the rules which Mr. Airy gave for adjusting compasses by magnets. Iron bolts, bars, or tubes, if placed in an upright position, in an upright ship, would continue in an upright position as the ship turned round on the water. The iron would remain at a constant angle with the *dip*, and exhibit a constant amount of magnetism as the ship was swung round, and would exert a permanent magnetic intensity on the compass, so long as the ship continued in the same latitude. Should the ship change her geographical position, the correcting bars would undergo the same kind of change as the iron in the vessel ; and if the correction did not actually remain perfect in all parts of the world, at all events, a deviation compass would shew any changes that might take place.[*] Permanent magnets, applied to correct variable disturbing forces, are now known to have double compass errors in some ships that have passed from one hemisphere to another.

§ 63. Although the accidental and unpremeditated arrangement, or stowage of some ships may have placed the metals in such quantities or relative positions, that the binnacle compass has exhibited no errors in an upright position ; it might turn out, that an inclination of the vessel might cause its compass to depart from the correct magnetic meridian. This kind of error may, when once ascertained by trial, be corrected by soft iron. A Pendulous apparatus (§ 54), although inconvenient, might accomplish that object, by being made to hang like a barometer, either over or under the compass, so that, as the ship inclined, the upper or lower end might pass from one side of the compass to the other, and exert a counteracting magnetism to that of the ship. This plan, although it appeared to afford pleasure and satisfaction to the Lords of the Admiralty, when they witnessed its success, is imperfect, and would be very unsatisfactory in stormy weather. The inclination errors and magnetic oscillations of the compass are easily got over, if the binnacle compass be conveniently placed to admit of *one* bar of iron being placed directly under it, when the north point of the compass-needle is drawn towards the lee side, as was the case in the *Recruit*.

On looking back to § 25, *figs.* 7 and 8, the reader will observe, that the magnetism of the guns drew the north point of the needle to leeward, and that opposite guns did not cancel errors, but doubled them. In the *Recruit*, her guns, when secured in their ports, acted magnetically to combine with her iron sides in causing incline deviations in the same direction. If *a beam* had been directly *under* the *Recruit's* compass, to which an iron bar of three or four feet in length could have been secured, with its *centre* directly under the compass,[†] the polarity of the bar would have passed from end to end,

[*] The deviation compass needle passes through twice the arc of the deviation of the main needle, and for that reason alone it can only be used on board ship in smooth water. W.W.

[†] It has been said, that if a mass of iron be directly under a compass, its disturbing force on the compass will vanish. " P vanishes if the mass of iron be exactly below the compass." This is a mistake. Recollect that the *Recruit's* beams were all of wood.

as the ship rolled or inclined (like an air bubble in a spirit level), the highest end of the bar counteracting exactly the magnetism of the lee side of the ship, and *vice versâ.*

There are, however, some iron ships with iron beams, in which the north point of the compass is drawn to *windward* instead of to leeward; and in such vessels two transverse bars of iron would be required for correcting or cancelling inclination deviations : for let an horizontal bar be placed athwartships on each side of the compass, it will be evident to the reader, that such an arrangement would produce an effect directly opposite to that of a single bar, with its centre directly under a compass.

In order to adjust a ship's compass, its errors must first be ascertained in an upright position, on each point, with everything in its place secured for sea service; care being taken that a line be drawn from " lubber's point " in the compass bowl to the pivot of the compass-card be parallel to the ship's keel. Let the ship's head be placed in the direction of the correct magnetic meridian, and if the north point be attached to starboard or port, let this deviation (which is generally very small), be corrected by a bar or bolt of soft iron, to be secured in a vertical position abreast of the compass, and not too close to it. The errors at north and south being cancelled, let the ship's head be placed in an east and west direction, where the errors are generally greatest. These errors may be corrected by another but larger bar, bolt, or tube* of soft iron, placed in a vertical position, either directly before or abaft the centre of the compass-card. Should the deviations be easterly, with the ship's head at east, the correcting iron should be fixed abaft the compass, with its upper end on a level with the plane of the compass-card; or, if more convenient and practicable, the iron might be placed in a vertical position before the compass, with its lower end on a level with the plane of the compass-card. One precaution is most essential, the iron should be large enough to control the compass-errors by the induced polarity it receives from the earth; for, if small, a powerful compass-needle would not be under sufficient control of the induced iron magnet. In all probability, pieces of iron about twice the weight and dimensions of the artificial magnets now used for correcting compasses would be sufficiently powerful. In iron ships, there are sometimes vertical pillars placed for supporting the decks : these *supports* might be turned to good account as correctors of compass deviations. This process is simple, inexpensive, and without risk; and should errors be found at N.E., S.E., S.W., or N.W., a little box of nails or an iron chain should be applied in the way recommended by Mr. Airy, and pursued by those who have been familiar with the process of adjusting ships' compasses by artificial " permanent magnets."

When the process above described has been gone through, the ship should be heeled by water casks, coals, stones, sand, or anything not magnetic; and her keel placed in the direction of the *correct* magnetic meridian. If an iron vessel, it will be found that

* A cylindrical tube of given dimensions will act magnetically as powerful as a solid bolt of the same dimensions and description of metal.

the compass is considerably in error, without the application of one
or two bars of soft iron, placed athwartships in a direction *parallel
to the ship's beam*, and directly abreast of the centre of the compass.
Two bars will be required, when it is found that the north point of
the compass-needle has been attracted by the weather or highest
side of the ship in her inclined position; and a single bar, with its
centre directly under the compass, when the attraction of the vessel
is the other way. These bars being skilfully placed, so as to make
the compass correct with the ship's head at north or south, with an
inclination say = 8° to each side; then by swinging the ship round
to east or west, the compass will be found sensibly correct. If the
ship's head be brought back again to north or south, and held steadily
in the direction of the magnetic meridian, the ship may be gradually trimmed to an upright position; and if during this operation
the compass be found correct, no further process will be necessary.

We have said, that ships' compasses are sometimes made correct
by accident; it is surely desirable to accomplish our object by intention, and by means of the same kind of materials which derange the
free action of a ship's compass. A deviation compass shews in
smooth water those parts in ships where there are no errors, and we
have endeavoured to shew, that we may obtain an equation, or zero
point of no disturbance, by an addition or transposition of the metals
in a vessel, with almost as much certainty as an algebraist forms an
equation, when he clearly understands the nature of his proposition,
or thing to be worked out.

§ 64. I have now before me about half a score of printed pamphlets
on great circle sailing and compass deviations, in wood and in iron
ships, containing rules and methods for correcting compass errors,
and instructions for navigators for making the quickest passages
across the ocean. The authors of these productions have been
generally without much practical knowledge of navigation, or experience at sea; and their works have tended more to bewilder than
enlighten mariners of ordinary capacities and common sense. It is
probably, on this account, that the Lords of the Admiralty have
simply ordered that magnetism in ships and compass deviations be
inserted in the *Examination Sheets* at the Royal Naval College,
Portsmouth, and that the magnetism of ships' and the mariners'
compass, should be a subject of instruction in the navigation school
of the college at Greenwich.

The superior officers of the marine department of the Board of
Trade, have likewise inserted in the Examination papers, required
under the mercantile marine act, a problem or explanation, requiring masters under examination to explain their views of compass
deviations; how it may be found, and how it should be applied in
correcting a ship's compass course, and those masters who volun-
tarily offer themselves for voluntary examination for the display of
superior abilities, are also required to explain the nature of great
circle sailing without going into the calculations thereof. Here I
may mention (parenthetically), that Nautical Astronomy, Education,
and Theoretical Navigation at Liverpool suddenly suffered a terrible
collapse; for during the five years that voluntary examination of

masters and mates went on, and when certificates of competency were issued by the local examiners, *assisted* by *teachers*, almost every one examined at Liverpool obtained a first class certificate, or an extra first class certificate! Strange to say, no sooner had the act passed for compulsory examinations, when it was found that the mariners of Liverpool were very much on a par with the other "Mariners of England." The first Official Mercantile Marine Navy Lists exhibit the superiority of the Liverpool masters and mates during the above mentioned short period, and the sudden change that took place in 1850.

As regards the advantage derivable from the knowledge of great circle sailing, it amounts just to this,—when the compass bearing of the port of destination is daily inserted in the ship's log book, by mercator's sailing and *also by great circle sailing*, it is at once seen how the ship should be steered when the wind is fair; and on what tack she should stand when the wind is not fair; for everybody knows that a ship should take the nearest road to her place of destination when other circumstances admit her to do so. Now the great circle road is shorter than the mercator's road across the ocean; the last is curved, the first is "as the crow flies."

Having commented abundantly on the inductive magnetism of iron and iron ships, and mentioned the different drugs that compass doctors and pamphleteers would prescribe for the cure of compass errors, and rectification of tables of deviation, when ships pass from one magnetic latitude to another, I shall conclude with a few remarks on the necessity of navigators giving their best attention to the amplitudes and azimuths they are required to solve on their examination sheets, before they obtain their certificates of competency. As a practical navigator of some experience, I never required a table of compass deviations for general use, and I still regard the numerous brochures, circulated gratuitously, or sold dearly, as likely to puzzle seamen of ordinary abilities, and divert them from more important matters contained in their "Epitome's," such as that of the lamented "Raper," whose book contains almost every thing that a navigator requires.

1st. On every clear day at sea, and from sun rise to sun set, three celestial observations may be made to determine the error or correction that the mariners' compass may require for correcting the course the ship is steering at the time of observation, and without any computation at all.

The sun's *observed* amplitude at sun rise or sun set, (when the binnacle or standard compass is used,) has only to be compared with the sun's true amplitude to be taken out of a "table of amplitudes" opposite to the ship's latitude in, and the sun's declination for the time of observation; of course the difference between the sun's true bearing and his compass bearing, at rising or setting, is the *correction* for the course on which the ship is *steering* at the instant of observation; and in the correction so easily obtained, we always have the variation properly called, and the deviation caused by the ship's local attraction combined: whether the ship be perfectly upright or heeling under canvas.

When the sun is on the meridian at noon, he bears N. or S.; and the difference between the sun's *true* bearing and his compass bearing, is always the compass correction, for the point the ship's head is on, when the sun is on the meridian.

2nd. An altitude of the sun, taken to find the longitude by chronometer, and the azimuth or sun's compass bearing, may be computed simultaneously, and by the same data, and at the same opening of the *Epitome*; the apparent time at ship (hour angle), and sun's true azimuth, may be obtained in the usual daily process at sea of "taking sights," that is to say, the longitude by chronometer, and also the correction of the compass course is obtained at the *time* of observation, whether the ship be perfectly upright or inclining under canvas, and found correctly by a single process!

Azimuths for compass errors and "sights" for the chronometers should be taken two or three hours before noon, or after noon, and I am now to prove that a single observation may answer for a double purpose, with very little labour, whenever the sun's compass bearing is taken, and his altitude is taken for obtaining his hour angle, and consequent time from noon, for determining the ship's longitude by chronometer.

Example:—A ship being in latitude 50° 0′ N., when the sun's declination was 20° 0′ S. Observed the true altitude of the sun, 14° 0′. The bearing of the sun's centre being noted at the same time. Required the apparent time from noon, and the sun's true azimuth.

SOLUTIONS.

FOR HOUR ANGLE.				FOR AZIMUTH.			
	° ′				° ′		
Altitude	14 0			Polar distance	110 0		
Latitude	50 0	Sec.	0.191933	Latitude	50 0	Sec.	0.191933
Polar dist.	110 0	Cosecant	0.027014	Altitude	14 0	Sec.	0.013005
Sum	174 0			Sum	174 0		
Half Sum	87 0	Cos.	8.178800	Half Sum	87 0	Cos.	8.718800
Remainder	73 0	Sin.	9.980596	Difference	23 0	Cos.	9.964282
	h. m. s.				° ′ ″		
Hr. angle } appt. time} at ship }	2 13 50	Sine Squared	8.918343	True Azimuth. }	32 18 30	Sine Sqd.	8.887365
	9 46 10						

Here it is manifest that a single observation answers a double purpose. The same data in an inverted order, enables us to obtain the longitude by the chronometer, and the actual correction for the compass course the ship is steering at the time, the sun's compass bearing is taken.

If the above observation had been made in the chops of our channel, near Scilly, where the variation is known to be 25° westerly, and if the ship was a wooden one, without any local attraction, the sun's compass bearing in altitude would have been found S 7° 18′ 30″ E. at 9 46 10 a.m.

If the observation had been made in the iron sloop of war *Recruit*, which had been swung in three positions, (sec. 58) where the errors of her compass are recorded, the sun's compass bearings would have been exceeding complicated, and would have puzzled even a "Philadelphia Lawyer instead of Captain Slade and the officers of the brig." The iron sided and wooden beamed brig would have had a new error for every course and every degree of inclination under sail, but an azimuth computed as above would have given the compass correction on any course whatever.

As the *Recruit's* deviations are inserted in (sec. 58), a tabular form, those who may have any doubts about the above results, may make the calculations and satisfy themselves. Although the commander of the Recruit was *ordered* to use and be guided by the *upright deviation* table furnished by the Superintendent of Her Majesty's Compass Department, it was not without sufficient cause he had no confidence in it, after being at sea in the brig in a dense fog in the English channel.

I have said that I look upon the numerous pamphlets on compass deviations with dis-favour; they are, generally speaking, not needed by those who know how to find their compass errors by celestial observations. The masters of our coasters and of our home trade passenger ships, although good seamen, good pilots, and prudent men, are generally not navigators.

In closing my remarks, I will mention that Mr. Burdwood, master, R.N., of the Hydrographic department at the Admiralty, has computed tables of the sun's true bearing in altitude, which might prove of much use to masters of coasters, and of home trade ships, because they might, by means of the tables, get the sun's true bearing from latitude 49 to 52 north, and by taking the sun's compass bearings, obtain the error of their compass courses. It should, however, be borne in mind that the time required for using the tables, should be the sea or sun time, and not Greenwich time, now so commonly in use on town clocks and pocket watches.

Mr. Burdwood will be pleased to see that the example of azimuth I have worked out on assumed data, agrees exactly with the result obtained by using his table for latitude 50°, and the time from noon as found in the example I have given.

§ 65. The officers of our mercantile marine are now required to understand the nature of a ship's local attraction; how it may be found; and when found they should know how to correct their reckonings by it. If young men, desirous of being proficients in the art of navigation, will carefully read over the remarks we have roughly strung together for their benefit, and make the experiments we have described, with a ship's or boat's compass, and a piece of iron out of the carpenter's chest or locker, they will thereby acquire useful information relative to the errors of compasses; and, also, as to how cargoes of iron should be stowed, in order to lessen the dangers arising from carrying ship loads of iron or machinery.

A few practical hints may be profitably placed in a condensed form, and in such language as may be understood by seamen generally; or by any one who has read the foregoing essay, and acquired some practical knowledge of keeping a ship's reckoning at sea: I mean that *dead reckoning* which is dependent upon recorded compass courses, log-line distances, and judgment allowances, carefully made for each course on the log slate, for lee-way, lulls of wind, heave of the sea, or bad steering. It is in this that the art of navigation consists. An expert schoolmaster might easily find the position of a ship, when the sun, moon, stars, or planets are visible; but to keep good dead reckonings for two or three days, or to shape correct courses, and steer clear of danger in dark nights, fog, and drizzle, is quite another thing.

The experience of half a century has firmly convinced me, that sufficient attention is not paid to the elementary principles of a common day's work. I have found gentlemen proud of knowing how to clear a lunar distance, or work a double altitude; and yet knew not how to correct the compass course for lee-way, deviation, and variation; and who never saw an azimuth compass in a merchant ship.

The following practical hints are thrown out for the benefit of mariners generally:—

1st. Since one powerful compass-needle, at a distance of two feet from another of the same kind, will produce an error in each compass of four degrees, it is therefore prudent and proper to have only one compass in a ship's binnacle.

2nd. In order to avoid unequal attractions of iron quarter davits, iron knees, and bolts at the sides of the ship, let the binnacle compass be placed on a midship fore and aft line.

3rd. See that the lubber's point in your compass-bowl, and the upright pivot for the card, are in a line with the ship's stem and stern-post; and prove this by stretching a thread over them, in an exact fore and aft direction.

4th. Take care that no iron of any kind be near your compass, either on the deck or under it; for arms are sometimes stowed below, and cabin stoves and funnels, with iron linings, are sometimes too near to the steering compass.

5th. The binnacle compass being the helmsman's guide, it is seldom convenient for taking bearings of distant objects. Compasses placed in different parts of the same ship, point in different directions, and have different errors for the same course: consequently a bearing, amplitude, or azimuth taken on a ship's gangway or forecastle, may be very different from observations made on the quarter deck. It is, therefore, necessary that all bearings taken, observations made, and courses shaped and recorded, should be made by a good instrument, at one convenient place on a ship's upper deck, on amidship fore and aft line, where the deviations are small, or equal to nothing; or where it may be convenient to correct a compass by iron alone. A compass of this kind is called a standard compass, by reason of its permanent position. It may be fixed on the top of a permanent pillar, or on a tripod made to ship and unship, for greater convenience.

6th. The standard compass, when in its proper place, should be high enough to admit of bearings being taken, and amplitudes observed, over the sides of the vessel. The compass-needle should be powerful and yet steady in a sea-way. The card should be an azimuth card, with a deviation apparatus over it, to indicate any errors that might be induced by unforeseen circumstances. The deviation style facilitates the taking of bearings; and by its shadow the sun's amplitude, or bearings in azimuth, are conveniently read off on the compass-card.

7th. When a ship is steering a given course, and an amplitude or azimuth is taken and worked out, we find the difference between the sun's *true* bearing, and his bearing by the compass we used at the time. Now this difference we call the variation, when in point of fact the ship's local attraction is involved with it.

What we really obtain, is a *correction* to be applied to *the course the ship is steering by the compass we use* in making our observations; and the result we obtain may be the sum or difference of the compass deviation on that course, and the variation properly so called in that particular place.

All that the navigator has to do, is to apply the result of his observations to correct the course steered at the time, *as if it were variation, and nothing else.**

8th. The magnetism of a ship and the errors of her compass undergo changes as she passes from one latitude to another; increasing or diminishing with the dip, and changing in character and quantity as the ship passes out of one magnetic hemisphere into another: and since iron takes up magnetism in less time than it parts with it, a portion of *time* will be required to develope fully the compass deviations of a ship, after moving rapidly from one magnetic dip to another.

9th. Since the magnetism of ships and the errors of their compasses are found to be *variable quantities*, as the ships move from place to place; to permanently correct these variable quantities, by means of magnetic steel bars, assumed to possess permanent magnetism, appears to be impossible. But to apply, for the purpose of correction, the same kind of metals to correct, as the metals that disturb or derange, is not inconsistent with inductive reasoning, experience, or common sense.

10th. Prudence and precaution require, that mariners should use all available means for determining the errors of their compass at sea, or in an anchorage; and since these means are generally attainable, we shall notice some of them,

11th. Whenever a ship anchors, or is becalmed, where some fixed object is easily seen at a distance of eight or ten miles, its compass bearing should be noted, as well as the direction of the ship's head. Write down the points of the compass, and as the ship swings by wind or tide from one point to another, write down the compass bearing of the distant object, opposite to the direction of the ship's head. If the compass be in error, the bearings of the distant object will vary, as the ship swings from point to point. As the ship swings round, there will be two nearly opposite points of the compass on which the bearings of the distant object agree; and these two points give the correct magnetic bearing of the object, and the points upon which there is no compass error. Two other points will be found, in a table so filled up—and nearly opposite points—where the

* In the olden times, before variation charts were drawn, and compass deviations demonstrated, azimuth compasses were in general use: but now there is not one merchant ship in fifty that carries an azimuth compass. How then can observations be made? Far better would it be that ships were supplied with three or four good instruments, than with half a dozen erroneous and imperfect compasses, requiring constant repair and re-touching, and in volving endless expense.

bearings of the distant object greatly differ! Now the differences between the correct magnetic bearing, and the incorrect bearings on other compass points will give the deviation on these points.

This is an easy and pretty correct way of finding a ship's compass deviations at anchor, when a ship is supplied with a proper compass for the purpose. The true bearing of a distant object may be found by measuring, with a quadrant, its angular distance with the rising or setting sun. The sun's correct bearing, at rising or setting, is at once obtained from a table of amplitudes; and hence the true bearing of any visible object is easily found.

12th. When a ship is at sea, we have shewn that, in amplitude and azimuth calculations, the variation and deviations are involved. When the sun is on the meridian, he bears true north or south, and a compass bearing of him at that time will give the variation and deviation combined, and consequently the *correction* for the ship's course at the time.

In taking a sight for the chronometer, the sun's compass bearing should be noted; as the *sight* might be used for finding the *hour angle and azimuth* together. Because in both solutions, the Latitude, Altitude, and Polar Distance are required, and by solving both problems by the same opening of the book, the hour angle and Longitude is found, and also the sun's bearing in azimuth and compass bearing give the compass error on the course steered at the time of observation, and whether the vessel be upright or inclining under sail. We have given an example of the process.

In our hemisphere, the North Pole star bears nearly due north; and its compass bearing will, therefore, give the variation and deviations combined.

There is another method whereby we may easily find out, at sea, if our binnacle compass be in error. In beating to windward, with northerly or southerly winds, it sometimes appears that the ship, after tacking, must either have been off the wind, or else a change of wind, has taken place: " She lay W. ¼ S, on the other tack, and now she is only E. ¼ S." On other occasions, in working to windward, the ship appears to be within 4½ points of the wind. All this indicates compass deviation. The wind had not changed; it was the compass pointing that changed, as the ship tacked from west to east.

13th. In iron sailing vessels, the compass errors are very great near the ends and sides of the ships, and least of all on a fore and aft midship line, near the middle of the vessels' length. When an iron sailing ship's head is at east, a compass near the stern almost always exhibits *westerly* deviation;* whilst another compass placed forward would shew an *easterly* deviation; both ends of sharp ship's attracting the north point of the compass on easterly and westerly courses. In such vessels, the standard compass should be fixed between these extremes, where the deviation is inconsiderable, or may be reduced by artificial means. The standard compass should be on the deviation compass principle, in order to indicate changes, should any

* Westerly deviation in north magnetic latitude, and easterly deviation by a compass near the bow.

take place by inclination or otherwise; and also for conveniently taking amplitudes or azimuths.

14th. To an iron steam-ship's hull, there is super-added iron engines, boilers, and funnel, fixed near the middle of the vessel. Now, the magnetism of the iron *ends* of the ship is counteracted to some extent, by the machinery and middle parts of the vessel; and somewhere between the stern-post and funnel, and on a fore and aft line on the upper deck, a place may be found where the compass deviation is *nil*, or nearly so. There a standard compass should be fixed. In iron steamers, a compass may be placed further aft than in an iron sailing vessel of the same form and magnitude.

In H.M. iron steamer *Dover*, the deviation of her standard compass at west was only 5° 30' E. It would have been less if the compass had been a little further forward.

In H.M. iron steamer *Trident*, the deviation at west was 17° W.; and in this ship, a compass nearer the stern would have had less deviation—or even no deviation at all.

In H.M. late iron steamer *Birkenhead*, the deviation of her standard compass at west was 12° 10' *easterly*; and here the stern and after parts of the ship had the magnetic mastery over the engines, boilers, funnel, and everything else. If the *Birkenhead's* compass had been placed further forward, her deviations would have been reduced, or cancelled altogether.*

From the foregoing detail of " facts and figures," we may conclude, that in all ships whatever, there may be found a position for a standard compass *better than any other;* and that with a little skill, a large share of common sense, and practical experience, such a position may be found.

* The Birkenhead ran on a sunken (Sec. 58.) rock, off Point Danger, (discovered by the Author,) the iron ship broke directly across, in the direction of a water-tight Bulkhead, and in the direction of the rivet perforations made from side to side. Her iron frame broke like postage stamp paper, and exactly on the same principle. Nearly all hands perished. W. W.

H. V. HARRIS, Printer to Her Majesty, Devonport.